The Image of Restoration Science

This book is about a single image – the frontispiece to Thomas Sprat's *History of the Royal Society of London* (1667). Designed by John Evelyn and etched by Wenceslaus Hollar, this is arguably the best-known representation of seventeenth-century English science. It is an outstanding example of the use of images to celebrate and legitimise the 'new' science, in which historians have become increasingly interested. Nevertheless, many questions arise as to its history and its rationale. Was it an original composition by Evelyn, or did it have earlier exemplars? What significance should be attached to all the instruments, books and other objects that appear in it? And how did it come to be designed in the first place?

In fact, although published as the frontispiece to Sprat's *History*, the print had originally been intended for a different work that proved abortive, and the book opens by examining the implications of this for Evelyn's design. The rest of the volume provides a full analysis of Evelyn's image in its Royal Society setting. It considers the print's debt to earlier exemplars, showing how closely its design is modelled on the frontispiece to a work celebrating Raphael, thus suggesting that Evelyn may have been trying to align the society as much with the legacy of Raphael as that of Copernicus. The book then examines the myriad details included in the print, particularly the society's institutional accoutrements and the scientific instruments that are profusely depicted, which closely reflect the inventions and experiments associated with the society at the time the etching was made. The book concludes by considering the print's publishing history, including evidence that choice copies were produced on *gampi* paper from Japan, thus exemplifying the combination of technological and artistic accomplishment to which Evelyn believed the society should aspire.

Michael Hunter FBA is Emeritus Professor of History at Birkbeck College, University of London. He is the author of numerous books, including *Boyle: Between God and Science* (2009) and *Boyle Studies: Aspects of the Life and Thought of Robert Boyle (1627–91)* (2015), and he is the principal editor of Boyle's *Works* and *Correspondence* (1999–2001). He has also edited *Printed Images in Early Modern Britain: Essays in Interpretation* (2010).

Jim Bennett is Keeper Emeritus of the Science Museum, London, and former Director of the Museum of the History of Science, University of Oxford. He is a leading expert on scientific instruments of the period.

Figure 0.1 The frontispiece to Sprat's *History of the Royal Society*.

The Image of Restoration Science

The frontispiece to Thomas Sprat's
History of the Royal Society (1667)

Michael Hunter
With a chapter on the instruments by Jim Bennett

Routledge
Taylor & Francis Group

LONDON AND NEW YORK

First published 2017
by Routledge
2 Park Square, Milton Park, Abingdon, Oxon OX14 4RN

and by Routledge
605 Third Avenue, New York, NY 10017

First issued in paperback 2022

Routledge is an imprint of the Taylor & Francis Group, an informa business

Publisher's Note
The publisher has gone to great lengths to ensure the quality of this reprint but
points out that some imperfections in the original copies may be apparent.

British Library Cataloguing in Publication Data
A catalogue record for this book is available from the British Library

Library of Congress Cataloging in Publication Data
Names: Hunter, Michael, 1949– | Bennett, J. A. (James A.)
Title: The image of restoration science : the frontispiece to Thomas
    Sprat's History of the Royal Society (1667) / by Michael Hunter ;
    with a chapter on the instruments / by Jim Bennett.
Description: London ; New York : Routledge, 2017. | Includes index.
Identifiers: LCCN 2016029584 | ISBN 9781472478726 (hardback : alk. paper) |
    ISBN 9781315556857 (ebook)
Subjects: LCSH: Scientific illustration—England—London—History—17th
    century. | Scientific illustration—Great Britain—History—17th century. |
    Science—Great Britain—History—17th century. | Sprat, Thomas, 1635–1713.
    History of the Royal-society of London, for the improving of natural
    knowledge. | Frontispiece.
Classification: LCC Q222 .H8625 2017 | DDC 506/.042—dc23
LC record available at https://lccn.loc.gov/2016029584

ISBN 13: 978-1-4724-7872-6 (hbk)
ISBN 13: 978-1-03-240227-7 (pbk)
ISBN 13: 978-1-315-55685-7 (ebk)

DOI: 10.4324/9781315556857

Typeset in Sabon
by Apex CoVantage, LLC

This book is dedicated to the memory of
Michael Cooper (1935–2012)
and
Lisa Jardine (1944–2015),
co-authors of
London's Leonardo: The Life and Work of Robert Hooke
(Oxford, 2003),
by its surviving co-authors.

Contents

Illustrations

Note: Figures 4.1b, 4.2a, 4.4, 4.5, 4.6b, 4.7, 4.9a, 5.1, 5.2, 5.5, 5.6, 5.8, 5.10, 5.11, 5.13, 5.17, 5.20, 5.22, 5.24, 5.28, 5.31, 5.32, 5.35, 5.36, 5.37, 5.38, and 5.39 are taken from the 'presentation' copy of the frontispiece at the Royal Society (RCN 7384): see Figure 6.1.

Acknowledgements

This has been a truly collaborative study, to which many have contributed and in the preparation of which important debts have been accumulated. The most significant collaboration has been between Michael Hunter and Jim Bennett: although it was the former who initiated the enterprise, building on the appendix devoted to the frontispiece that was published in his *Science and Society in Restoration England* (Cambridge, 1981), the latter, as the leading expert on instruments of the kind that festoon the background to the print, has been heavily involved almost from the outset. After initially offering advice on the identification of the instruments and establishing contact with other experts in the field, it then became apparent that the scale of treatment that was required of this aspect of the print would be best served by his writing a whole chapter himself on the instruments that are shown in it. This therefore appears as chapter 5. Here, it is perhaps worth noting that both this and the remainder of the text have been the subject of mutual discussion between the two authors, so some ideas that originated with Michael Hunter appear in chapter 5, while suggestions by Jim Bennett have enhanced the remainder of the text.

Particular thanks must be recorded to Nathan Flis of the Yale Center for British Art, New Haven, who first pointed out the overriding debt of Evelyn's design to Nicolas Chaperon (the link had initially been missed due to reliance on the 1675 Aquila/Fantetti/Maratti edition of the Raphael Bible rather than Chaperon's 1649 one!). Nathan has also read a draft meticulously and made other very useful suggestions. Others who have read a draft of the work and supplied helpful comments on it are Roger Gaskell, Robert Harding, Sachiko Kusukawa, Margaret Lock, Scott Mandelbrote, Anna Marie Roos, Ad Stijnman, Ben Thomas, Anthony Turner and Simon Turner.

Many of these have also answered specific queries and provided other forms of assistance, as have the following (in alphabetical order, to avoid any invidious ranking between them): David Alexander, Peter Anstey, Rupert Baker, Domenico Bertoloni Meli, Jonathan Betts, Ann Blair, Elizabethanne Boran, The Lord Camoys, John Cheetham, Andrew Crawforth, Diana Crawforth-Hitchins, Judith Curthoys, Gillian Darley, Howard Dawes, Mark Dawson, Richard Dunn, Patrick Evelyn, Jonathan Ferguson,

David Galbraith, Antony Griffiths, Frances Harris, David Helliwell, Felicity Henderson, Matthew Hunter, Andrea Immel, Stephen Johnston, Malcolm Jones, Suzanne Karr Schmidt, Vera Keller, Roger Keynes, John Lawson, Fred Lock, Giles Mandelbrote, Katherine Marshall, David McKitterick, Joanna McManus, Christopher Mendez, Jenny Millar, Megumi Mizumura, Keith Moore, Andrew Nahum, Sheila O'Connell, Keith Piggott, Lawrence M. Principe, Graeme Rimer, Martin Rudwick, Nicholas Stogdon, Stuart Talbot, John Taylor, Philip Trower, Alena Volrábová, Jonathan White, Frances Willmoth and Huib Zuidervaart. We would also like to thank the following, whose assistance has often gone far beyond the call of duty: Elma Brenner, Susanna Cimmino, Alan Coates, Julie Cochrane, Carly Collier, Sophie Defrance, Rembrandt Duits, Susan Halpert, Andrew Honey, Rebecca Horton, Jane Hughes, Kathryn James, Berthold Kress, Elenor Ling, AnnaLee Pauls, Adam Perkins, Mary Peskett Smith, Angela Roche, Nikolai Serikoff, Joanna Snelling, Kevin Spears and Grant Young. Jon Wilson has provided invaluable help in editing the illustrations and getting them into publishable form. We are also grateful to Tom Gray for recognising the book's potential and accepting it for publication, and to Michael Bourne for seeing it through the Ashgate/Routledge transition. At Routledge, Kitty Imbert, Jo Hardern and Max Novick have expertly supervised the production process.

A generous Publications Grant covering digitisation costs and reproduction fees has been made by the Paul Mellon Centre for Studies in British Art.

Lastly, oral presentations on this topic were given by Michael Hunter at the Royal Society on 28 March 2014; at Indiana University, Bloomington, on 19 September 2014 (when it formed the R.S. Westfall Lecture); at the University of Sydney on 13 April 2015; at Stanford University on 6 October 2015; and at the Institute for Research in the Humanities, University of Bucharest, on 7 April 2016. He is indebted to his hosts or hostesses on those occasions: Rupert Baker, Domenico Bertoloni Meli, Peter Anstey, Paula Findlen and Dana Jalobeanu. He is also grateful to all who attended for the valuable comments that they made in the discussion that followed, many of which have been taken into account in preparing the text of the book.

Abbreviations

The following abbreviations are repeatedly used in the notes:

Add. MS	British Library Additional Manuscript
bpi1700	British Printed Images to 1700 website, http://www.bpi1700.org.uk/index.html
Birch, *Royal Society*	Thomas Birch, *The History of the Royal Society of London* (4 vols., London, 1756–7)
Boyle, *Correspondence*	Michael Hunter, Antonio Clericuzio and Lawrence M. Principe (eds), *The Correspondence of Robert Boyle* (6 vols., London, 2001)
Boyle, *Works*	Michael Hunter and Edward B. Davis (eds), *The Works of Robert Boyle* (14 vols., London, 1999–2000)
Cl. P.	Royal Society Classified Papers
Evelyn, *Diary*	E. S. de Beer (ed.), *The Diary of John Evelyn* (6 vols., Oxford, 1955)
Grew, *Musæum*	Nehemiah Grew, *Musæum Regalis Societatis* (London, 1681)
Hunter, *Establishing*	Michael Hunter, *Establishing the New Science: The Experience of the Early Royal Society* (Woodbridge, 1989)
Keynes, *John Evelyn*	Sir Geoffrey Keynes, *John Evelyn: A Study in Bibliophily with a Bibliography of his Writings* (2nd edition, Oxford, 1968)
New Hollstein	Simon Turner, edited by Guilia Bartram, *The New Hollstein: German Engravings, Etchings and Woodcuts 1400–1700: Wenceslaus Hollar,* parts 1–9 (Ouderkerk aan den IJssel, 2009–12)
Notes & Records	*Notes & Records of the Royal Society of London*
Oldenburg	A. R. and M. B. Hall (eds), *The Correspondence of Henry Oldenburg* (13 vols., Madison, Milwaukee and London, 1965–86)

Pennington, *Hollar*	Richard Pennington, *A Descriptive Catalogue of the Etched Work of Wenceslaus Hollar* (Cambridge, 1982)
Phil. Trans	*Philosophical Transactions*
RBO	Royal Society Original Register Book
Sprat, *History*	Thomas Sprat, *The History of the Royal Society of London* (London, 1667)

Quotations from manuscript sources are presented according to the principles expounded in Boyle, *Correspondence*, vol. 1, pp. xli–xlii, and Boyle, *Works*, vol. 1, p. cii. Briefly, original spelling, capitalisation and punctuation are retained, whereas standard contractions (e.g., the thorn with superscript 'e' for 'the') have been silently expanded. Underlining in the original is shown by the use of italics. Editorial insertions have been denoted by square brackets. Deletions and words inserted above the line in the original are recorded in endnotes.

1   Introduction

This book is a study of a single image – though arguably one of the best known that has come down to us from seventeenth-century science. This is the plate that was published as the frontispiece to Thomas Sprat's *History of the Royal Society of London* (1667). It was designed by the virtuoso and diarist, John Evelyn, and etched by Wenceslaus Hollar, the Bohemian artist long domiciled in England.[1] Within an architectural setting surmounted by the Royal Society's coat of arms and festooned with scientifically significant books and objects, it shows a bust of King Charles II, the society's founder and patron, placed on a pedestal and being crowned with a wreath by the goddess Fame. The column on which Charles's bust stands is flanked to the right by the society's first President, William, Viscount Brouncker, and to the left by Francis Bacon, Lord Verulam, the figure from the early seventeenth century who more than anyone else was the society's intellectual inspiration. Bacon provided a mandate for the complete reformulation of knowledge about the natural world, not least through the collection of a great inductive 'natural history', and he also emphasised the desirability of the application of science to practical use. Here, he is shown in his robes as Lord Chancellor.

In seeking to encapsulate the message of a book in a prefatory image, the print that precedes Sprat's *History* falls into a tradition of providing frontispieces and engraved title-pages that by the late seventeenth century was well established both in Britain and in Europe more generally.[2] Indeed, the use of such plates to celebrate and legitimise the 'new' science of the period has recently begun to attract attention from historians.[3] In the words of the Jesuit astronomer, Christoph Scheiner, 'a small image teaches what many written words cannot say,' and this was a point that was not lost on authors like Galileo, whose works were embellished with striking illustrations, as with the title-page that Stefano Della Bella designed for the *Dialogue concerning the Two Chief World Systems* (1632), which powerfully evokes the message of Galileo's book (Fig. 1.1).[4]

Indeed, through such images, astronomers like Tycho Brahe and Andrea Argoli sought to establish their lineage from Atlas and Hercules and to raise the status of their discipline accordingly, while a succession of authors used

Figure 1.1 Stefano Della Bella's etched title-page for Galileo's *Dialogue concerning the Two Chief World Systems* (1632), showing Aristotle, Ptolemy and Copernicus.

elaborate designs – often including personifications based on biblical and classical imagery – as a means to dignify the pursuits in which they were engaged. By comparison, the Sprat frontispiece might seem rather artful and understated in the way in which it puts across its message, but it has nevertheless been reproduced very frequently. This is largely because it is the most striking image relating to the early Royal Society that we have, encapsulating the society's ethos in significant ways that will be explored in the course of this study.[5] Indeed, no book on the early Royal Society and its context seems complete without a reproduction of it.

It is also worth noting the context of Sprat's *History of the Royal Society* itself in terms of self-promoting publications by the new institutions devoted to natural philosophy of which the Royal Society was one. Thus the Accademia del Cimento, founded under the auspices of the Medici family at Florence in 1657, brought out a volume of *Saggi di Naturali Esperienze* which, though not actually published until 1667, had been in preparation since the start of the decade.[6] Equally notable were the publications of the Académie des Sciences founded in Paris in 1666, including the grandiose volumes devoted to *L'Histoire des Animaux*, originally published in 1671, and *L'Histoire des Plantes*, published in 1676, elephant folios that were distributed under the auspices of the French monarchy.[7] Both the volumes just mentioned feature lavish etchings by Sébastien Leclerc which provide a profuse pictorial record of the academy's members and patrons, and of its apparatus, activities and milieu (Figs. 1.2, 5.21).[8] We entirely lack an English equivalent to this mouth-watering visual resource, and it is perhaps partly for this reason that the Evelyn/Hollar image has come to acquire such prominence in relation to the seventeenth-century Royal Society.

Nevertheless, there are many questions to be asked about it and how it came into being. Was it an original composition by Evelyn, or is it based on earlier exemplars? Can one identify all the instruments, books and other objects that appear in it, and what significance should be attached to their inclusion? Perhaps above all, how did the plate come to be designed in the first place, and what is its true relationship with Sprat's book? The latter question is a pressing one, since the status of the print in relation to the volume is problematic. Facing the title-page of the book, on the verso of the half-title, is a depiction of the society's coat of arms (see Fig. 2.1); in copies of the book in which the etched image is to be found, it is often inserted rather uncomfortably between that and the title-page. In addition, many copies of the book lack the Evelyn/Hollar plate, although these are often in contemporary bindings and there is no sign that it has been removed. In any case, the print is larger in size than the format of normal copies of the book, which means that, insofar as it appears in these, it has had to be cropped or folded to make it fit. It *is* of just about an appropriate size for the choice copies of the book that were printed on large paper, and it has sometimes been claimed that it must have been produced for inclusion in these: but this seems to be a retrospective rationalisation.[9] Instead, there is a clear reason for the

Figure 1.2 Sébastien Leclerc's etching of Louis XIV's (imagined) visit to the Académie des Sciences in Paris, from *Mémoires pour Servir à l'Histoire Naturelle des Animaux* (1671).

anomalous status of the plate, which was first divulged in 1981 and will be recapitulated here. This is that it was not originally intended for Sprat's work at all. In fact, it was designed for a completely different book, a defence of the Royal Society devised by the Somerset virtuoso, John Beale, and it was re-routed to Sprat's volume only when Beale's project proved abortive.[10]

This book will therefore begin by giving an account of Beale's scheme and its context, prior to providing a full analysis of Evelyn's image in its Royal Society setting. This will comprise three parts. First, we will consider the overall iconography of the image and its – in some respects surprising – message concerning Evelyn's conception of the society's role. We will then examine the myriad of details included in the plate and their significance, looking first particularly at the portraits and books included in it and the trappings of the society's institutional status. A further chapter will then give minute attention to the various instruments that are depicted. Such scrutiny of the print's details raises questions about its design process, including the relative contribution of designer and etcher, and these are addressed in the context of seventeenth-century print production more generally. In a final chapter, we will consider the print's history after its publication, including the extent to which Evelyn seems to have used copies of it to exemplify the combination of technological and artistic accomplishment to which he believed the Royal Society should aspire.

Notes

1 The print is described and briefly commented on in virtually all modern accounts of Hollar and of Evelyn, as follows: Howard C. Levis, *Extracts from the Diaries and Correspondence of John Evelyn and Samuel Pepys Relating to Engraving* (London, 1915), pp. 140–2; F. G. Grossmann, *Wenceslaus Hollar 1607–77: Drawings, Paintings and Etchings* (Manchester, 1963), no. E19; Keynes, *John Evelyn*, pp. 283–4; Katherine S. van Eerde, *Wenceslaus Hollar: Delineator of His Time* (Charlottesville, 1970), pp. 80–1; Graham Parry, *Hollar's England: A Mid-Seventeenth-Century View* (Salisbury, 1980), p. 29 and plate 107; Pennington, *Hollar*, no. 459; Antony Griffiths and Gabriela Kesnerová, *Wenceslaus Hollar: Prints and Drawings* (London, 1983), no. 110 (p. 69); Richard T. Godfrey, *Wenceslaus Hollar: A Bohemian Artist in England* (New Haven and London, 1994), p. 22; Frances Harris and Michael Hunter (eds), *John Evelyn and His Milieu* (London, 2003), p. 16; Anne Thackray, *Caterpillars and Cathedrals: The Art of Wenceslaus Hollar* (Toronto, 2010), pp. 90–1; *New Hollstein*, no. 1966; Douglas D. C. Chambers and David Galbraith (eds), *The Letterbooks of John Evelyn* (2 vols., Toronto, 2014), vol. 1, p. 438. Where not purely descriptive, most accounts of the print are positive, but for an exception see Gillian Darley, *John Evelyn: Living for Ingenuity* (New Haven and London, 2006), p. 186, who speaks of 'the dull image which Evelyn and Beale designed' (see also p. 212). Two further useful works on Hollar which do not, however, mention the Sprat frontispiece are Jacqueline Burgers, *Wenceslaus Hollar: Seventeenth-Century Prints from the Museum Boymans-van Beuningen, Rotterdam* (Alexandria, VA, 1994), and Rachel Doggett, Julie L. Biggs and Carol Brobeck, *Impressions of Wenceslaus Hollar* (Washington, DC, 1996).

2 For English examples see Margery Corbett and Ronald Lightbown, *The Comely Frontispiece: The Emblematic Title-Page in England 1550–1660* (London, 1979) and A. F. Johnson, *A Catalogue of Engraved and Etched English Title-Pages Down to the Death of William Faithorne, 1691* (London, 1934). For continental examples see next note.

3 See esp. Volker R. Remmert, *Picturing the Scientific Revolution* (Philadelphia, 2011), translated by Ben Kern from Remmert, *Widmung, Welterklärung und Wissenschaftslegitimierung* (Wiesbaden, 2005).

4 Ibid., pp. 57, 59–60, 212.

5 For engraved title-pages associated with the society, see that to Richard Waller's translation of the *Saggi* of the Accademia del Cimento (London, 1684) and that to Francis Willughby's *Historia Piscium* (Oxford, 1686). These are more typically allegorical in the tradition studied by Remmert. It is perhaps revealing that the latter was the most expensive plate in the book, costing £4, far more than the others. See Sachiko Kusukawa, 'The *Historia Piscium* (1686)', *Notes & Records*, 54 (2000), 179–97, esp. p. 191. See also her 'Picturing Knowledge in the Early Royal Society: The Examples of Richard Waller and Henry Hunt', *Notes & Records*, 65 (2011), 273–94 (where the *Saggi* title-page is reproduced and discussed), and Johnson, *Catalogue*, pp. 53, 77.

6 See W. E. Knowles Middleton, *The Experimenters. A Study of the Accademia del Cimento* (Baltimore, 1971), pp. 65ff. and passim, and Alfonso Mirto, 'Genesis of the *Saggi* and Its Publishing Success in the Seventeenth through Nineteenth Centuries', in Marco Beretta, Antonio Clericuzio and Lawrence M. Principe (eds), *The Accademia del Cimento and Its European Context* (Sagamore Beach, 2009), pp. 135–49.

7 Anita Guerrini, 'The King's Animals and the King's Books: The Illustrations for the Paris Academy's *Histoire des Animaux*', *Annals of Science*, 67 (2010), 383–404. See also Roger Hahn, *The Anatomy of a Scientific Institution: The Paris Academy of Sciences, 1666–1803* (Berkeley and Los Angeles, 1971), Alice Stroup, *A Company of Scientists: Botany, Patronage and Community in the Seventeenth-Century Parisian Royal Academy of Sciences* (Berkeley and Los Angeles, 1990) and Eric Brian and Christiane Demeulenaere-Douyère (eds), *Histoire et Mémoire de l'Académie des Sciences* (Paris, 1996), esp. pp. 107ff.

8 See E. C. Watson, 'The Early Days of the Académie des Sciences as Portrayed in the Engravings of Sébastien Le Clerc', *Osiris*, 7 (1939), 556–87, and Maxime Préaud, '"L'Académie des Sciences et des Beaux-arts": Le Testament Graphique de Sébastien Leclerc', *RACAR: Revue d'Art Canadienne*, 10 (1983), 73–81. The most striking of the engravings are reproduced, e.g., in Hahn, *Anatomy of a Scientific Institution*, p. 1, fig. 2, and endpapers, and Stroup, *Company of Scientists*, pp. 6, 40 and 42.

9 Keynes, *John Evelyn*, p. 284, citing the views of H. B. Wheatley, which 'I have not been able to verify'. This claim appears in H. B. Wheatley, *The Early History of the Royal Society* (Hertford, 1905), p. 2 (a paper read at a meeting of the Sette of Odd Volumes in 1894). It also appears in a MS note by Wheatley dated 1901 in a copy of the book formerly owned by him that was later owned by Keynes and is now in the possession of a member of his family: this contains the plate, although is not a large paper copy (see Fig. 6.3), thus explaining the scepticism with which Wheatley's claim that it is such a copy is repeated in Geoffrey Keynes, *Bibliotheca Bibliographici* (London, 1964), p. 367; it is perhaps also worth noting that this is one of the few books from Keynes's collection as listed in *Bibliotheca Bibliographici* which did not go to Cambridge University Library in 1982. The claim that the plate is only found in large paper copies also appears in Jackson I. Cope

and Harold Whitmore Jones (eds), *The History of the Royal Society by Thomas Sprat* (London, 1959), p. ix. However, although it seems likely that all large-paper copies originally contained the frontispiece, copies of it are also found bound into regular-sized copies of the book in early bindings: see chapter 6.

10 Michael Hunter, *Science and Society in Restoration England* (Cambridge, 1981), pp. 194–7.

2 John Beale, 'Lord Bacon's Elogyes' and the fortunes of Sprat's *History*

Vindicating the Royal Society

In considering the plate's context, we need to go back to the beginnings of the Royal Society itself. It is easy to take the society's foundation in 1660 for granted as a natural corollary of the growing esteem for science in mid-seventeenth-century England, and it is certainly true that science had thrived during the Interregnum and that it continued to flourish in more or less formal milieux thereafter.[1] It is equally true that a galaxy of famous names was associated with the society throughout the late seventeenth century, including Robert Boyle, Robert Hooke, William Petty, Christopher Wren, John Flamsteed, Isaac Newton and many others. But the inauguration of the Royal Society was significant because it reflected a view that the advancement of science would benefit from its being represented by a public institution – in other words, from its being 'established', to which end the infant body was given an elaborate constitutional structure involving statutes and a charter (which was replaced by a revised one within a year of being issued). Indeed, the society represented a new type of public body, the prototype of all the voluntary institutions devoted to learned ends that have been founded ever since.[2]

Yet, insofar as its structure and functions lacked precedent, the corollary was that the society's early years saw a prolonged period of experimentation, as its founders discovered just what it was feasible for such a body to achieve, and what it was not.[3] At the outset, the society had breathtaking ambitions for the scale and comprehensiveness of the reform of knowledge about the natural world that, inspired by Bacon, it hoped to achieve. Moreover it was initially intended that these plans would be realised by corporate activity at its meetings: 'it was to be a kind of once-a-week Rockefeller Institute', as Robert G. Frank tellingly put it in a pioneering study.[4] Yet such aspirations proved to be over-ambitious or impractical. Increasingly, the society's experimental programme became dependent on the heroic figure of its Curator of Experiments, Robert Hooke, appointed to the post in 1662, who will bulk large in this study.[5] Meanwhile its correspondence was the responsibility of the first Secretary, Henry Oldenburg, who proved assiduous in keeping in touch with scientific enthusiasts at home and abroad.[6] By such means the society gradually evolved into a type of body that had not at first been entirely anticipated, in which its predominant role was less in the actual performance of science than in the provision of a clearing house for ideas about all aspects of the natural world and an agency for evaluating, accrediting and disseminating them – functions in which it was to prove remarkably effective.

The society's novelty also meant that it was much misunderstood by contemporaries, many of whom regarded it with considerable suspicion, if not outright hostility; others subjected it to satire, while some failed to take it seriously at all. It was reported in 1664 that even the King, despite being its official patron, 'mightily laughed at [the society] for spending time only in weighing of ayre, and doing nothing else since they sat'.[7] The leading figures of the society took the view that such attitudes were at least partly due to the fact that the society's objectives were not properly understood, and for this reason the idea was early formulated of producing a kind of *apologia* for the society, explaining its aims and the potential benefits that might derive from its activities, and reassuring any who might feel threatened by it.[8] As early as June 1662 a committee was appointed 'to draw up a paper concerning the design of the society', while in the following May Sir Robert Moray, a courtier and leading Fellow, mentioned in a letter to his Dutch correspondent, Christiaan Huygens, the idea of publishing 'a little treatise, by which you can learn all that concerns the society'.[9] It was evidently to this end that Thomas Sprat was elected to the society in April 1663. Sprat was a graduate of Wadham College, Oxford, and a protégé of one of the leading figures in the society in its early years, John Wilkins, former Warden of the college and now Dean of Ripon. More importantly, Sprat was a budding *litterateur*, a member of the circle of court wits that also included authors like Abraham Cowley and John Dryden, and he was clearly appointed with the specific task of deploying his 'knack of fair speaking' in composing the book that was ultimately to be published in 1667 (Fig. 2.1).[10]

However, all did not go smoothly. Sprat does not seem to have worked on the task as quickly as Wilkins and others had hoped, and it appears that, when nothing had materialised after over a year, the society had to start to put pressure on him to complete his assignment. Finally, in November 1664, perhaps thanks to influence exerted through his patron, the Duke of Buckingham, Sprat delivered quite a significant amount of the text of the book – perhaps as much as a third of the volume as it was to appear in 1667 – and there were even hopes that it might shortly be ready to go to press.[11] But matters were then delayed while the draft was examined by a Royal Society committee, whose concern seems to have been especially with the 'particulars' of the society's work that Sprat would include, in other words, details of the investigations that the society had carried out, the apparatus that it had developed, and the like. Further delay was caused when the Great Plague struck London in the summer of 1665, which led to meetings

Figure 2.1 Title-page of the *History of the Royal Society* that Thomas Sprat wrote between 1663 and 1667, faced by the coat of arms granted to the society in its second charter, a typical facet of its early institutional ambitions.

of the society being intermitted for several months, and nothing had come to fruition by the following September, when the Fire of London consumed much of the city; this necessitated the society's move from Gresham College, the educational institute in the City where it had initially been established, which was commandeered in the aftermath of the Fire for civic purposes. All this further distracted both Sprat and the society.[12]

John Beale and his project

As a result of these developments, for some years it might well have seemed as if the work was not going to materialise at all. This must explain why an alternative apologetic initiative emerged, which is crucial here because it was almost certainly for this that the plate that became the frontispiece to Sprat's volume was originally intended. The rival project was the brainchild, not of the society's organisers at its London base, but of one of its leading provincial supporters, the Somerset divine and virtuoso, John Beale. Beale is a fascinating figure, a passionate enthusiast for the new science and for intellectual reform more generally who must have written more long letters on related topics than almost anyone else of his period.[13] Educated at Eton College and

Figure 2.2 Francis Bacon, whose advocacy of intellectual reform was inspirational both to Beale and to the Royal Society as a whole. Engraved portrait by Simon de Passe, 1618.

King's College, Cambridge, of which he became a Fellow, in the 1630s Beale spent two years travelling abroad as a tutor. He then had a rather troubled experience during the Civil War and its aftermath, resigning his Fellowship of King's and being forced to take refuge in his native Herefordshire. In the 1650s he became a devoted associate of the intelligencer, Samuel Hartlib, with whom he corresponded on many topics until Hartlib's death in 1662. It was with Hartlib's encouragement that Beale published his *Herefordshire Orchards, A Pattern for all England* in 1657, and he continued to promote horticulture and cider-making thereafter, not least following his move to Somerset, where he was Rector of Yeovil from soon after the Restoration until his death in 1683.

Throughout this period, Beale corresponded with Robert Boyle, John Evelyn and Henry Oldenburg (in the latter case, until Oldenburg's death in 1677): he initially came into contact with these men through Hartlib, and he wrote as assiduously to all of them as he had previously to Hartlib himself. Though rarely in London, Beale was elected a Fellow of the Royal Society in January 1663, and he took a devoted interest in the society and its fortunes from his Somerset home. He was one of those most actively involved in the society's initiative to promote silviculture that materialised as Evelyn's *Sylva, or a Discourse of Forest-Trees*, published with the society's imprimatur in 1664; a substantial section of this, entitled *Pomona*, was devoted to cider and prominent in this was the contribution of Beale.[14] This formed part of Beale's attempt to interest the society in a plan to promote cider-making all over England, which he had submitted through Oldenburg in November or December 1662.[15]

In 1664, Beale seems to have been the stimulus behind one of the society's most significant early institutional initiatives, the setting up of eight committees dealing with each of the principal areas of its interests – though it was symptomatic of the society's over-ambitiousness in its early years that none of these bodies met for very long.[16] In March that year Beale offered 'to communicate several observations on agriculture, if the society should please to appoint a committee to receive and examine them', and as result 'it was thought good, that the council should consider of this and other committees for several subjects.'[17] The remit of the committees ranged from 'collecting the phaenomena of nature and all philosophicall experiments hitherto observed, made and recorded' to 'correspondence', which mainly involved the compilation of 'inquiries' to be sent to exotic places – a characteristically Baconian initiative that formed one of the society's most significant activities in its early years.[18] Perhaps the most active was the 'Georgical' or agricultural committee, the direct result of Beale's initiative, which assiduously collected information about husbandry and related topics.[19] Second only to this was the 'Mechanical Committee', which dealt with areas where it was hoped that the society might produce initiatives of immediate utility, reaching a climax in March 1665 when a patent was obtained for a series of such inventions, ranging from new carriage designs to an engine for breaking flax and hemp and washing linen.[20] Since some of the inventions involved related to apparatus that appears in the Sprat frontispiece, we will return to them in chapter 5.

As for Beale, no less significant was his interest in issues concerning the publishing and promotion of science. In 1666 he sent a series of lengthy letters to Robert Boyle in which he advocated a collected edition of Boyle's writings as exemplary of the Royal Society's design, considering the order in which they should be presented and the format in which they would be most effective, though nothing came of this.[21] It is also hardly surprising that he was an early enthusiast for *Philosophical Transactions*, the pioneering scientific periodical launched by Henry Oldenburg in 1665, to which Beale became a profuse contributor, even advising Oldenburg on the wording of the dedication to volume 4 of the journal in 1670.[22] For Beale, such enterprises formed part of a broader strategy for what he called 'the Mastery of Fame', or 'the Management of fame'.[23] In this, he was inspired both by Francis Bacon himself, whose fragmentary essay 'Of Fame' had first been published in 1657, and also by Sir Henry Wotton, the former diplomat who had been Provost of Eton when Beale was at school there, and who seems to have become something of a mentor to the younger man.[24] Beale took their message to be for the need for vigorous propaganda to spread information and right opinion as widely as possible, using all appropriate means, especially 'the cheap rate and spreading advantage of Typography'.[25] Such a strategy was 'the moderne Lightning', as he presciently put it, 'Nothing more potent . . . yet nothing more neglected', and he reiterated this constantly in his letters to his various correspondents.[26] Indeed, this theme was to recur in the 1670s, when he saw that the widely circulating annual almanacs offered potential for promoting the new science that more staid publications lacked, and he continued assiduously to advocate the promotion of agricultural improvement by all available means – including the new periodical press – up to his death.[27]

The idea of a publication in defence of the Royal Society was a natural corollary to such preoccupations on Beale's part. His letters to his London contacts are full of complaints about the apathy and hostility that the society met and the need to overcome this. 'O the villanes', he exclaimed of the universities, 'who defame us as enemyes to Aristotles excellencyes, whereas they do only licke the foame of his animosityes, & the darke dregs of his lurking holes'; he was equally hostile to the royal court and to other members of the establishment whom he saw as apathetic to the society's aims.[28] It was apparently in November 1664 that Beale initially put forward the idea of a publication aimed at raising the society's profile among such groups, in other words just before Sprat belatedly produced his first instalment, and at a time when it may have seemed likely that he was never going to produce anything at all. From a letter from Beale to Evelyn which is undated but evidently of that month, we learn that he had 'treated & transacted lately' on the subject with Sir Robert Moray, who, as we have seen, was one of the society's leading figures at this point; he also referred to his 'notes' on the subject that were by then in the possession of Oldenburg.[29] The latter must have comprised the 'roll of papers, sent to me by Mr Beale about the asserting and establishing the reputation of the Society, both at home and abroad', which Oldenburg reported to Robert Boyle in a letter dated 17 November that he had been 'called upon to consider with some Company' that day.[30] Evidently those in high places in the society encouraged Beale in the plan that he outlined in these and subsequent letters, despite the fact that at this point Sprat's project was at last showing signs of coming to fruition: perhaps their feeling was that two initiatives might be better than one.

In any case, Beale's original plan was for something quite different from the lengthy quarto volume that Sprat was to produce. Throughout, he took it for granted that 'our way to support our owne Enterprise is to devise all wayes to revive Lord Bacons lustre', thus echoing the significance of Bacon for the society that has already been stressed.[31] But what he initially had in mind was a broadsheet, headed by an engraved portrait of Bacon, shown in his chancellor's robes ('a stately philosophicall habite'), which was to be accompanied by sections of text, possibly in the form of aphorisms from Bacon's writings. These might perhaps have been separated by classical columns, recalling those representing the Pillars of Hercules that appear on the title-page of Bacon's *Novum Organum* (1620), and the idea was that they might be engraved, possibly in a variety of elaborate scripts ('so many severall characters') by Edward Cocker, one of the leading calligraphers of the day (Fig. 2.3).[32]

One possible analogue that Beale might have had in mind were the broadsheets of approximately foolscap format that were widely distributed in seventeenth-century England, often reaching sections of the market that

Figure 2.3 Edward Cocker, the calligrapher who Beale thought might execute his engraved broadsheet, but who did not feel up to the task.

Figure 2.4 A typical example of a Jesuit thesis print, that of the Polish Jesuit, Gabriel Kilian de Bobrek Ligeza, to be defended in Douai on 22 August 1628. This vast sheet (measuring 919 by 595 mm), by Schelte Adamsz. Bolswert after Peter Paul Rubens and Pieter Soutman, shows Sigismund III, King of Poland and Sweden, enthroned in an elaborate architectural setting; under him appear the propositions to be defended, surrounded by allegorical scenes. Beale's projected broadsheet might possibly have borne some resemblance to exemplars like this.

books did not, and this may have been part of his plan.[33] Alternatively, he might have had in mind something more ambitious, like the much larger Jesuit thesis prints that were produced in other parts of Europe at this time, which combined elaborate decorative designs with an engraved or letterpress text comprising a list of propositions to be defended (in this instance, these would have been replaced by Baconian precepts).[34] Beale's knowledge of more than merely English exemplars is suggested by the fact that, in a letter to Evelyn of 22 April 1665, he mentioned that the print and bookseller, Roger Daniel, who had printed Beale's *Herefordshire Orchards*, had told Beale's cousin, Sir John Pye, that 'He knewe one in Fr[ance] or Italy that got £8000 sterling by such a devise.'[35] That he had in mind something on a large scale is also implied by the fact that Cocker, the calligrapher whom Beale had earmarked for the task, evidently considered himself inadequate to it.[36]

Beale continued actively to plan his promotional work until 1667, but his conception of it seems to have evolved somewhat, largely due to the difficulties he encountered in trying to realise his initial idea of using calligraphy as the medium for his publication.[37] He continued to refer to it by the same title, 'Lord Bacon's Elogyes &c', since he remained convinced that the promotion of Bacon was the best means to gain support for the Royal Society.[38] Indeed, throughout this period and beyond he advocated making Bacon's writings more widely available as 'the likeliest expedient, to Angle for the young Lawyers, who are the generality of the English Gentry', whose support he saw as crucial for the success of the Royal Society's programme.[39] Instead of a broadsheet, however – whether majestic or popular – he seems to have come to think more in terms of a printed book. This would have comprised (in a formatted heading which Beale provided in an undated letter to Evelyn, probably of early 1667):

A briefe Representation in sumary heades
pointing out The Progresse of Solid Learning
Since my Lord Bacon invited the Ingenuous
To addict to faythfull experiments
And more especially since it was countenanced
And undertaken by the Royall Society.

It was to be 'a volumne rather too large, than too briefe for his Majestyes perusall', and various details of it are divulged in Beale's letters, not least to Evelyn.[40] Indeed, at this point Beale even tried to increase Evelyn's enthusiasm for the project by drafting a preface to the book, in which readers were urged to study Bacon's works and to learn more about the extent to which his vision had since been realised, which was to be signed by Evelyn's son, John Evelyn junior, then just starting his studies at Trinity College, Oxford, under his tutor, Ralph Bohun.[41]

Evelyn's design

More significantly, Beale also enrolled Evelyn's help to provide an illustration for the book. As we will see in the next chapter, Evelyn was not only closely involved with the Royal Society but was also an artistic connoisseur and writer, and Beale was fully aware of this; indeed, his supportiveness for Evelyn's role in bringing European artistic precepts to an English audience is reflected in the dedicatory verses that he provided for Evelyn's translation of Roland Fréart, Sieur de Chambray's, *Parallel of the Antient Architecture with the Modern*, published in 1664.[42] Beale must also have known that Evelyn was a draughtsman in his own right, who had produced more than one set of etchings in the years around 1650.[43] Hence his suggestion that Evelyn might devise a plate for his book was not surprising. Evelyn evidently agreed to produce a design for Beale, and it is clear that this came to fruition while Beale's volume was still being actively planned. Indeed, certain details in the print as it was to materialise echo aspects of Beale's views as expressed in his letters.

For one thing, the idea of including the figure of Fame in the composition would have seemed natural in the light of Beale's preoccupation with fame and its 'mastery'.[44] Otherwise, his clearest statements as to the components that he saw as desirable were made in a letter to Evelyn of 22 April 1665, when Beale was wavering in his enthusiasm for an engraved broadsheet and was instead moving towards the idea of a printed book. The letter pursues the idea that the design might include columns of the main architectural orders to symbolise each of the three Stuart monarchs, Doric for 'the pacific James', Ionic for 'our Modest Augustus' (i.e., Charles I), and 'all the trimmings of the Corinthian perfection' for 'our Magnificent Founder'. He also wrote:

> I had halfe a wish, That Lord Bacon had beene drawne sitting under one Arch in his Chancellors roabes, which bore some resemblance of a philosopher civilised & in latter under another Arch adioyned might be his Apotheosis, the two Angells &c. But Sir I can make noe pretence at all to theese designes. I shall value it as a very greate favour if you shall please to direct according to your owne iudgement.[45]

In addition, it is possible that Beale may have had some initiative in suggesting certain of the instruments that were to be depicted in the print. Thus he was fascinated by barometers, providing barometric readings that were published in *Philosophical Transactions* in the 1660s and writing enthusiastically about the instrument as 'one of the most wonderful that ever was in the World', in that 'the seasons and changes of the Air and Weather can be thereby discover'd, and the now hidden causes of many other *Phænomena* detected.'[46] He was also interested in thermometers, in January 1667 devising a scheme which he sent to Robert Boyle for a bizarre spiral thermometer for use in greenhouses (Fig. 2.5).[47]

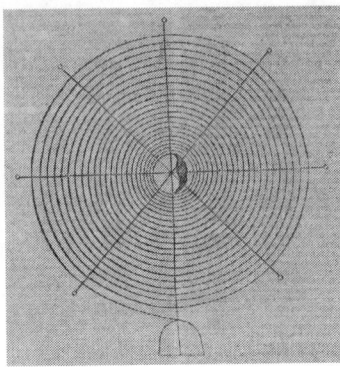

Figure 2.5 John Beale's idea for a spiral thermometer, as sent in a letter to Robert Boyle in January 1667.

The demise of Beale's scheme

In the end, Beale's book was never published. Though he was still working on it early in 1667, in April that year he heard that Sprat's book was imminent. The position was that, after the disruption caused by the Plague and Fire, by this time Sprat had returned to work on the project. It seems likely that it was at this point that the sections of the volume that he had written previously – namely, Part I, surveying the history of ancient and modern philosophy, and the opening section of Part II, comprising a narrative of the society's origins and an account of its aims and constitution – were complemented by further sections to produce the work as we now have it.[48] Two components were evidently added at this stage. One was the remainder of Part II, comprising 'instances' of the society's work which were apparently selected and written up in the early months of 1667 with the help of members of the society, as is revealed by various references in its minutes.[49] The other was the lengthy Part III of the book, in which Sprat sought to defend the society's design for the benefit of a variety of interest groups, which he probably wrote almost entirely without supervision – something that was to prove rather embarrassing to the society when the work actually appeared and was digested by the reading public.[50]

The fresh burst of activity on the book in the spring of 1667 was evidently reported to Beale, and on 29 April he wrote to Evelyn: 'If Mr Sprats history be comeing abroad (of which I had secrete causes to despayre) it will expunge the best, the hearte, & allmost the Totall of my designes.' He then explained of his own project how 'I must lay this my error not only upon Mr Old[enburg] & Mr Hooke, but allso upon Honourable Mr Boyle, as I can shewe by expresse letters.'[51] Quite what the 'secrete causes' were which had made him doubt whether Sprat's book would materialise are unclear (they probably related to his understanding of the reception of Sprat's riposte to an account of England by the French savant, Samuel de Sorbière, published the previous year).[52] In any case, although Beale seems to have come to appreciate that Sprat's book *was* imminent, he seems to have remained uncertain for at least another month as to how completely his own would be 'expunged' by it, since in a letter to Evelyn of 20 May 1667 he noted: 'I shall nowe wayte for Mr Sprats History to informe me, Whether there be any roome left for my recomends.'[53] At some point, however, Beale seems to have decided to abandon his own book, instead evidently looking forward to the appearance of Sprat's. In a letter to Oldenburg of 1 June he wrote: 'I hope Mr Sprats Hist[ory] of R[oyal] So[ciety] is ready to looke abroad,' requesting Oldenburg to bespeak five or six copies 'well bound without needeless garnishes' for him from the stationer.[54] He evidently received these later in the summer, and in a letter to Evelyn dated 11 September he gave an almost ecstatic review of the volume and the parties whom he thought it had placated – from parliamentarians to academics – writing: 'I make no doubt but it will prove an Invincible Light to all the world, & for all ages.'[55] On the other hand, (like many others) he was within a few years to experience severe doubts about the advisability of the apologetic message that Sprat put across.[56]

Here, what matters is the fate of the illustration that Evelyn had evidently prepared for Beale's book prior to the demise of his project. This is explicitly dealt with in his letter of 29 April, since his comments about the extent to which his apologetic was likely to be eclipsed by Sprat's that were quoted in the previous paragraph were preceded by the following: 'Now for your most obliging Drought, I pray you Let me whisper in your eare my reasons, why I am afrayd to claime it.'[57] Hence it is clear that by this time Evelyn had produced a 'Drought' for Beale's use, while Beale's embarrassment at having encouraged Evelyn to execute a task that had apparently become superfluous is evident from the convoluted way in which he made his further suggestion in the following paragraph:

> But Sir, If you can pardone a strange confidence, which is not without the fervent respects of unfeigned gratitude, according to the old rules of the best friendship; I wish the plate fitted for Mr Spratts 4°, & soe imployed allso, but still I have an unquenchable ambition to continue my power & propriety in your intended favor upon occasions. Can you pardone this freedome. In your intentions towards me you oblige Mr Sprat, the R Soc, posterity & I must name Mr Glanvill allso, & explaine it in a story.[58]

Beale was thus quite explicit in his wish to see 'the plate fitted for Mr Spratts 4°', even though matters were slightly complicated by the 'story' that he promised about his Somerset neighbour, the divine and polemicist, Joseph Glanvill, which may be elucidated as follows. In the letter, Beale went on to divulge his knowledge of the work by Glanvill that was to materialise as a further polemic on behalf of the Royal Society, published in 1668 with the title *Plus Ultra: Or, the Progress and Advancement of Knowledge since the Days of Aristotle, In an Account of Some of the most Remarkable Late Improvements of Practical, Useful Learning*. At one point in his letter Beale wondered if the plate might be equally suited to that, though he also expounded a different idea for an appropriate illustration to Glanvill's work (and for the actual thrust of Glanvill's polemic); in any case, none appeared in *Plus Ultra* when it came out the following year.[59]

Hence, even if it was not as immediately apparent to Beale as to Evelyn and his Royal Society colleagues either that Sprat's book would appear or that it would totally eclipse Beale's own (as witness the comments in his letter of 20 May already quoted), his letter of 29 April could nevertheless have been seen as providing a mandate for the reassignment of the image that Evelyn had designed for Beale to Sprat's book in the event of that being published. This must therefore mean that, as soon as Evelyn and his colleagues were sure that Sprat's *History* was indeed coming out, it would have seemed appropriate to take Beale up on his offer that the print should accompany Sprat's book rather than his. On the other hand, exactly what progress had been made in producing an actual plate from Evelyn's design at this point is a little unclear. As we have seen, Beale already refers to it as a 'plate', and the fact that it would have to be 'fitted' to the quarto format of Sprat's book implies that Beale's projected volume would have been of larger – perhaps folio – size, to which the print as it materialised would indeed have been well-suited. Indeed, this explains various of the anomalies about the plate's relationship with Sprat's book that were indicated in chapter 1 and to which we will return in chapter 6.

However, Beale's letter of 29 April goes on to include the following further comment: 'I do earnestly disswade you from the trouble of sending the draught for my judgment, which is nothing. I would rather make suite for it, when it is graven, not for the copper but for the draught.'[60] From these words, one might deduce not only that he had not himself yet seen the image but also that the design had not yet been etched, despite the fact that earlier in the letter he had referred to a 'plate' that would have to be fitted to Sprat's quarto. Yet if this was the case, and if by this time all that had occurred was that Evelyn had done a drawing, one would have thought that the exemplar could have been scaled down in size to fit Sprat's book: this was a normal practice at the time, since it was the task of the etcher, in this case Hollar, to adjust the design that he had been given to suit the format in which it was to be printed.[61] It is therefore equally likely that by now the plate itself was already in hand. This issue cannot be resolved but, whether merely designed or actually executed, the print that Beale had inspired now took on a different role, as part of Sprat's *History*, and it is to this that we must turn. We may end this chapter, however, by noting Beale's request 'not for the copper but for the draught': it is a pleasing thought that Beale might have been the recipient of the original drawing – though, if he was, it does not appear to survive.[62]

Notes

1 For recent studies, see Charles Webster, *The Great Instauration: Science, Medicine and Reform, 1626–60* (2nd edition, New York and Bern, 2002); Michael Hunter, *Science and Society in Restoration England* (Cambridge, 1981); Steven Shapin and Simon Schaffer, *Leviathan and the Air-Pump: Hobbes, Boyle and the Experimental Life* (2nd edition, Princeton, 2011); Marie Boas Hall, *Promoting Experimental Learning: Experiment and the Royal Society 1660–1727* (Cambridge, 1991); Lisa Jardine, *Ingenious Pursuits: Building the Scientific Revolution* (London, 1999); William T. Lynch, *Solomon's Child: Method in the Early Royal Society of London* (Stanford, 2001); and John Gribbin, *The Fellowship: The Story of a Revolution* (London, 2005).

2 See Hunter, *Establishing*, esp. ch. 1.

3 See Michael Hunter, 'First Steps in Institutionalisation: The Role of the Royal Society of London', in Tore Frängsmyr (ed.), *Solomon's House Revisited: The Organisation and Institutionalisation of Science* (Canton, MA, 1990), pp. 13–30, reprinted in Hunter, *Science and the Shape of Orthodoxy: Intellectual Change in Late Seventeenth-Century Britain* (Woodbridge, 1995), pp. 120–34; and the briefer statement in Hunter, 'The Great Experiment', *History Today*, November 2010, pp. 34–40.

4 R. G. Frank Jr, 'Institutional Structure and Scientific Activity in the Early Royal Society', in *Proceedings of the 14th Congress of the History of Science, 1974* (4 vols., Tokyo, 1975), vol. 4, pp. 82–101, on p. 84.

5 Hooke has been the subject of much attention in the past thirty years, notably in Michael Hunter and Simon Schaffer (eds), *Robert Hooke: New Studies* (Woodbridge, 1989); Stephen Inwood, *The Man Who Knew Too Much: The Strange and Inventive Life of Robert Hooke 1635–1703* (London, 2002); Jim Bennett, Michael Cooper, Michael Hunter and Lisa Jardine, *London's Leonardo: The Life and Work of Robert Hooke* (Oxford, 2003); Michael Cooper, '*A More Beautiful City': Robert Hooke and the Rebuilding of London after the Great Fire* (Stroud, Gloucs., 2003); Lisa Jardine, *The Curious Life of Robert Hooke, the Man Who Measured London* (London, 2003); Allan Chapman, *England's Leonardo: Robert Hooke and the Seventeenth-Century Scientific Revolution* (Bristol, 2005); Michael Cooper and Michael Hunter (eds), *Robert Hooke: Tercentennial Studies* (Aldershot, 2006); and Matthew C. Hunter, *Wicked Intelligence: Visual Art and the Science of Experiment in Restoration London* (Chicago, 2013).

6 See *Oldenburg*, passim; Hunter, *Establishing*, ch. 7; and Marie Boas Hall, *Henry Oldenburg: Shaping the Royal Society* (Oxford, 2002).

7 R. C. Latham and W. Matthews (eds), *The Diary of Samuel Pepys* (11 vols., London, 1970–83), vol. 5, p. 33.

8 On Sprat's *History* see esp. Paul B. Wood, 'Methodology and Apologetics: Thomas Sprat's *History of the Royal Society*', *British Journal for the History of Science*, 13 (1980), 1–26; Michael Hunter, 'Latitudinarianism and the "Ideology" of the Royal Society: Thomas Sprat's *History of the Royal Society* (1667) Reconsidered', in Richard Kroll, Richard Ashcraft and Perez Zagorin (eds), *Philosophy, Science and Religion in England 1640–1700* (Cambridge, 1992), pp. 199–229, also available in Hunter, *Establishing*, ch. 2; Lynch, *Solomon's Child*, ch. 5; John Morgan, 'Religious Conventions and Science in the Early Restoration: Reformation and "Israel" in Thomas Sprat's *History of the Royal Society* (1667)', *British Journal for the History of Science*, 42 (2009), 321–44; Morgan, 'Science, England's "Interest" and Universal Monarchy: The Making of Thomas Sprat's *History of the Royal Society*', *History of Science*, 47 (2009), 27–54. However, it is worth pointing out here that, for all its sophisticated analysis of the book's arguments, Morgan's work is vitiated by the fact that he fails to confront Hunter's argument in 'Latitudinarianism and the "Ideology" of the Early Royal Society'

that little can be learned about the society's stance from the later sections of Sprat's book. Instead, his articles are based on the presumption that, whatever Sprat's personal outlook, his *History* can be taken as a meaningful statement of the society's ideological position. For older studies largely superseded by this recent literature see Hunter, *Establishing*, pp. 46–7.

9   Birch, *Royal Society*, vol. 1, p. 85; Jackson I. Cope and Harold Whitmore Jones (eds), *The History of the Royal Society by Thomas Sprat* (London, 1959), p. xiii.

10 Fairfax to Oldenburg, 28 September 1667, *Oldenburg*, vol. 3, p. 492. For Sprat himself, see Hans Aarsleff, 'Sprat, Thomas', in C. C. Gillispie (ed.), *Dictionary of Scientific Biography* (16 vols., New York, 1970–80), vol. 12, pp. 580–7, and John Morgan, 'Sprat, Thomas (*bap.* 1635–d. 1713)', *Oxford Dictionary of National Biography*, Oxford University Press, 2004, online edition January 2008, http://www.oxforddnb.com/view/article/26173 (accessed 27 November 2015).

11 For the intervention of Buckingham, see the letter from Wallis to Moray quoted in Wood, 'Methodology and Apologetics', p. 4; for the delivery of the first part of the text, see Hunter, *Establishing*, p. 51.

12 Hunter, *Establishing*, pp. 51–2.

13 For biographical accounts see Mayling Stubbs, 'John Beale, Philosophical Gardener of Herefordshire. Part I. Prelude to the Royal Society (1608–1663)', *Annals of Science*, 39 (1982), 463–89, and 'John Beale, Philosophical Gardener of Herefordshire. Part II. The Improvement of Agriculture and Trade in the Royal Society (1663–1683)', *Annals of Science*, 46 (1989), 323–63.

14 John Evelyn, *Pomona* (London, 1664), esp. pp. 1, 5 (unsigned leaf between pp. 20 and 21) and 21–9.

15 See Beale to Oldenburg, 12 and 21 December 1662, in *Oldenburg*, vol. 1, pp. 479ff. (Beale's first letter to Oldenburg of 20 September 1659 appears on pp. 314ff.); Birch, *Royal Society*, vol. 1, pp. 144ff., 172, 176–7, 179, 368.

16 See Michael Hunter, 'An Experiment in Corporate Enterprise: The Royal Society's Committees of 1663–5, with a Transcript of the Surviving Minutes of Their Meetings', in Hunter, *Establishing*, ch. 3.

17 Birch, *Royal Society*, vol. 1, pp. 402–3.

18 Hunter, *Establishing*, esp. pp. 91–2, 93–4, 104–5, 118ff. On inquiries, see also Hunter, 'Robert Boyle and the Early Royal Society: A Reciprocal Exchange in the Making of Baconian Science', *British Journal for the History of Science*, 40 (2007), 1–23, reprinted in Hunter, *Boyle Studies* (Farnham, 2015), ch. 3.

19 See R. V. Lennard, 'English Agriculture under Charles II: The Evidence of the Royal Society's "Enquiries"', *Economic History Review*, 4 (1932–4), 23–45; Hunter, *Establishing*, esp. pp. 84ff., 105ff.

20 Hunter, *Establishing*, pp. 87ff., 114ff.

21 See Beale to Boyle, 18 April, 13 and 30 July and 10 August 1666, in Boyle, *Correspondence*, vol. 3, pp. 137–40, 186–210. For commentaries see Boyle, *Works*, vol. 1, p. lxxxiv, and Richard Yeo, *Notebooks, English Virtuosi and Early Modern Science* (Chicago, 2014), esp. pp. 146ff.

22 See Stubbs, 'John Beale', Part II, pp. 329–31; *Oldenburg*, vol. 6, pp. 474–7.

23 Beale to Oldenburg, 1 June 1667, *Oldenburg*, vol. 3, p. 426; Beale to Evelyn, 13 December 1662, Add. MS 78683, fol. 55v.

24 For some of the more significant of Beale's statements on this topic see Beale to Evelyn, 13 December 1662, 24 August 1667, Add. MS 78683, fol. 55v, Add. MS 78312, fol. 62v; Beale to Oldenburg, 15 January 1663, 1 June 1667, *Oldenburg*, vol. 2, pp. 6–7, vol. 3, p. 426; and Beale to Boyle, 21 November 1663, 26 June 1682, Boyle, *Correspondence*, vol. 2, pp. 207–10, vol. 5, pp. 303–4. For Bacon's essay see Francis Bacon, *The Essayes or Counsels, Civill and Morall*, ed. Michael Kiernan, Oxford Francis Bacon, vol. 15 (Oxford, 1985), pp. 177–8. For Wotton's role in relation to Beale, see Stubbs, 'John Beale', Part I, pp. 468–71.

25 Cl. P. 2, no. 22, Oldenburg's transcript of Beale's 'Of Chalcography an humble motion'.

26 Beale to Oldenburg, 15 January 1663, *Oldenburg*, vol. 2, pp. 6–7; Beale to Evelyn, 24 August 1667, Add. MS 78312, fols. 62–3.

27 See various letters of Beale to Christopher Wase, esp. 19 October 1672, 19 February 1673 and ND, in Corpus Christi College, Oxford, MS 332, fols. 22, 24 and 28. Related themes are also explored, e.g., in Beale to Evelyn, 2 January, 18 December 1669, 13 June 1674, Add. MS 78312, fols. 92–3, 119–20, Add. MS 78313, fols. 74–5, in Beale's letters in *Oldenburg*, vol. 9, and in Beale to Boyle, 26 June 1682, Boyle, *Correspondence*, vol. 5, pp. 298ff. See also Stubbs, 'John Beale', Part II, passim.

28 Beale to Evelyn, 10 February 1666, Add. MS 78312, fols. 42–3 ('excellencyes' is an insertion). For his overall concern see *Oldenburg*, passim; Boyle, *Correspondence*, passim; and Add. MSS 78312–3, passim.

29 Beale to Evelyn, November 1664, Add. MS 78312, fol. 12.

30 Oldenburg to Boyle, 17 November 1664, Boyle, *Correspondence*, vol. 2, p. 404. On those who encouraged him, cf. Beale to Evelyn, 29 April 1667, Add. MS 78312, fol. 50v, quoted on p. 18 in this book.

31 Beale to Evelyn, November 1664, Add. MS 78312, fol. 12.

32 Beale to Evelyn, 22, 26 April 1665, Add. MS 78312, fols. 31, 34. For Cocker, see Sir Ambrose Heal, *The English Writing-Masters and Their Copy-Books 1570–1800* (Cambridge, 1931), pp. 33–6, 135–45, and Alexander Globe, *Peter Stent, London Printseller c. 1642–65* (Vancouver, 1985), pp. 20–1, 203.

33 See Globe, *Peter Stent*, esp. pp. 17–19, 28. For broadsheets, see Tessa Watt, *Cheap Print and Popular Piety 1550–1640* (Cambridge, 1991), esp. ch. 6 for early seventeenth-century examples of broadsheets similar to what Beale might have had in mind; Antony Griffiths, *The Print in Stuart Britain 1603–89* (London, 1998), esp. pp. 144ff., 280ff.; and Malcolm Jones, *The Print in Early Modern England: An Historical Oversight* (New Haven and London, 2010), passim, who illustrates how items from the period studied by Watt continued to circulate in the later seventeenth century and beyond.

34 See Louise Rice, 'Jesuit Thesis Prints and the Festive Academic Defence at the Collegio Romano', in J. W. O'Malley, G. A. Bailey, S. J. Harris and T. F. Kennedy (eds), *The Jesuits: Cultures, Sciences and the Arts 1540–1773* (Toronto, 1999), pp. 148–69, and various examples of the genre in the British Museum, e.g., 1858,0417.1291, 1928,0313.356 or 1951,0407.50. We are grateful to Robert Harding for this suggestion. In this connection, it is perhaps worth noting that Beale's vision was to see Bacon depicted 'beyond our English Modell', with overtones of Solomon's throne or of St Jerome's description of that of Apollonius (Beale to Evelyn, November 1664, Add. MS 78312, fol. 12). Also interesting is his concern about the Jesuits and their ability to 'manage fame at their pleasure' (Beale to Evelyn, 24 May 1666, Add. MS 78312, fol. 38–9) (this was in connection with Evelyn's 1666 translation of Pierre Nicole's *Les Pernicieuses Conséquences de la Nouvelle Hérésie des Jesuites* (1664)), and the extent to which 'by their single sheetes they catch him that runneth by' (Beale to Boyle, 10 August 1666 (enclosure), Boyle, *Correspondence*, vol. 3, p. 200).

35 Beale to Evelyn, 22 April 1665, Add. MS 78312, fol. 31 (on Daniel see Globe, *Peter Stent*, p. 213; on the Pye family, see Stubbs, 'John Beale', Part 1, pp. 465–6).

36 See esp. Beale to Evelyn, 22, 26 April 1665, Add. MS 78312, fols. 31, 33–4.

37 See Beale to Evelyn, 22, 26 April 1665, Add. MS 78312, fols. 31, 33–4.

38 Beale to Evelyn [1667], Add. MS 78312, fols. 44–5. Cf. his letter of Nov. 1664, fol. 12 ('the Baconian elogyes').

39 Beale to Evelyn, January 1671, Add. MS 78313, fols. 34–5.

40 Beale to Evelyn [1667], 22 April 1665, Add. MS 78312, fols. 44–5 ('of the' is deleted after 'Representation'), 31v.

41 Beale to Evelyn [1667], Add. MS 78312, fol. 44. Cf. Evelyn, *Diary*, vol. 3, p. 474.

42 John Evelyn (trans), Roland Fréart, Sieur de Chambray, *Parallel of the Antient Architecture with the Modern* (London, 1664), sig a1.

43 See p. 33 in this book.

44 There is no need to see a specifically Rosicrucian allusion in this figure, as suggested in Frances A. Yates, *The Rosicrucian Enlightenment* (London, 1972), pp. 191–2. A more extreme claim of a similar kind appears in Robert Lomas, *The Invisible College: The Royal Society, Freemasonry and the Birth of Modern Science* (London, 2002), p. 80, who sees the depiction of Bacon as appearing 'amidst a welter of Masonic symbolism'.

45 Beale to Evelyn, 22 April 1665, Add. MS 78312, fol. 31 ('adioyned' is inserted).

46 Beale, 'A Relation of Some *Mercurial* Observations, and Their Results', *Phil. Trans.*, 1 (1665–6), 153–9, on p. 155. Further reports by Beale appeared in various subsequent volumes of *Phil. Trans.*

47 Beale to Boyle, 28 January 1667, Boyle, *Correspondence*, vol. 3, pp. 285–6. The illustration there (and here) is reproduced from the edition of the letter in question in Thomas Birch's 1744 edition of Boyle since the original has since been lost.

48 Sprat specifically states on p. 120 of his *History* that this was the point that he had reached in his first bout of work on the book. It is possible that Beale was sent copies of some of these sections: see Hunter, *Establishing*, pp. 51–2.

49 See Birch, *Royal Society*, vol. 2, pp. 161, 171, 176, the last entry for 23 May when the initiative was left with Brouncker and Wilkins; see also p. 138 for a reference to 'reducing the extracts of the society's journal-books into a method' for Sprat (in January), and pp. 163, 169 concerning the statutes.

50 For a detailed account of the publication of the book and its aftermath, see Hunter, *Establishing*, pp. 52ff.

51 Beale to Evelyn, 29 April 1667, Add. MS 78312, fol. 50v.

52 See Beale to Evelyn [1667], Add. MS 78312, fol. 44v. For Sprat's attack on Sorbière, see esp. Morgan, 'Science, England's "Interest" and Universal Monarchy', pp. 28ff.

53 Beale to Evelyn, 20 May 1667, Add. MS 78312, fol. 53v.

54 Beale to Oldenburg, 1 June 1667, *Oldenburg*, vol. 3, pp. 427–8.

55 Beale to Evelyn, 11 September 1667, Add. MS 78312, fols. 66–7.

56 Beale to Evelyn, 11 June 1670 and 14 March 1671, Add. MS 78313, fols. 13, 43, from which quotations are given in Hunter, *Establishing*, pp. 62–3 (where the latter is misdated 1670).

57 Beale to Evelyn, 29 April 1667, Add. MS 78312, fol. 50v ('it' is an insertion).

58 Beale to Evelyn, 29 April 1667, Add. MS 78312, fol. 51.

59 Ibid. His alternative idea was of the angel appearing to Hagar, which reflected his hope to persuade Glanvill to focus on the divine origin of inventions (for a Hollar print on this theme, see Pennington, *Hollar*, no. 65 (1); *New Hollstein*, no. 2428). On Glanvill and the local context to the origins of *Plus Ultra* see Nicholas H. Steneck, '"The Ballad of Robert Crosse and Joseph Glanvill" and the Background to *Plus Ultra*', *British Journal of the History of Science*, 14 (1981), 59–74.

60 Beale to Evelyn, 29 April 1667, Add. MS 78312, fol. 51v.

61 For such re-scaling on Hollar's part, see e.g the fresh versions of his *Aesop* illustrations that he prepared in the early 1670s, which are re-etched on a smaller scale from the plates published in 1665: Pennington, *Hollar*, nos. 333–90; *New Hollstein*, nos. 1857–1913, 2266–2305.

62 Though Beale's MSS are apparently lost, various books that he owned are to be found in Wells Cathedral Library, having been bequeathed by his successor as Rector of Yeovil, Martin Strong. We therefore investigated whether the library there might contain his copy of Sprat's *History*, conceivably embellished with the original drawing for the frontispiece. Sadly, the only copy of Sprat in the library was one formerly owned by Ralph Bathurst, Dean of Wells (which lacks the frontispiece). We are grateful to Kevin Spears, Librarian of Wells Cathedral, for his assistance in this connection.

3 The overall design of the frontispiece

Sources and significance

Evelyn and the Royal Society

At the outset, it must be stressed that the origins of the print as detailed above should warn against seeking too precise a correlation between it and the actual content of Sprat's *History*. Nevertheless it has to be admitted that, whether it was intended for Sprat's book or for Beale's, Evelyn's design would have served much the same purpose – namely of celebrating the infant Royal Society, its aims and achievements, its royal patronage, and its established status. Either way, Evelyn was the obvious person to take responsibility for such a composition. He had been closely associated with the society from an early date: though he was not one of the twelve men who met after a lecture by Christopher Wren on 28 November 1660 to inaugurate the society, his name appears in the 'catalogue' drawn up on that occasion of those 'fit to be joined with them in their design', and he was proposed for election on 26 December that year.[1] His significance in relation to the society is apparent from the fact that he may have given the society its name, or at least that he was the first person to use in print the name by which it was subsequently to be known, 'The Royal Society' (previously it seems to have been expected to have some title more aligned with its function, such as 'Society of Philosophers'[2]). This occurred in the dedication to a book that Evelyn published in 1661, his translation of Gabriel Naudé's *Instructions concerning Erecting of a Library*, and he specifically notes in his famous *Diary* that the society offered him its 'Publique Thanks' for this.[3] In the dedication to the Earl of Clarendon, Evelyn spoke highly of the society, seeing it as the agent of 'that glorious Work of Restoring the *Sciences*, Interpreting *Nature* [and] Inventing, and Augmenting of new and useful Things', as advocated by Bacon.[4] Indeed, John Beale considered these prefatory remarks one of the key texts in defence of the Royal Society, which he thought should receive wide circulation, 'For more may be sayd, but nothing can be better sayd'.[5]

It is also relevant that Evelyn seems to have been closely involved with the choice of a coat of arms and motto for the society, to judge from a sheet of paper in his hand now among the Evelyn Papers at the British Library endorsed 'Armes & Mottos proposd for the R[oyal] Society' (Fig. 3.2).[6] This shows not only the rather plain coat of arms that the society was to adopt – a blank shield quartered by the royal arms – but a variety of more

Figure 3.1 John Evelyn: pencil portrait by the French artist, Robert Nanteuil, 1650.

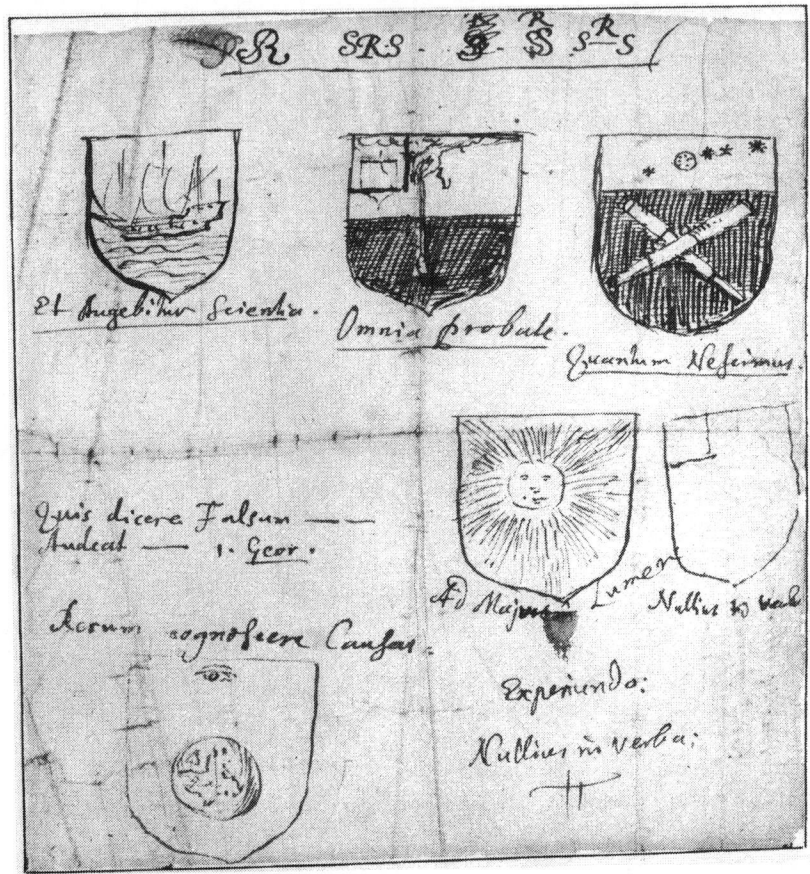

Figure 3.2 Evelyn's paper showing 'Armes & Mottos proposd for the R. Society'.

emblematic alternatives, including a ship under sail, a pair of crossed telescopes, a celestial globe under an all-seeing eye, a plumb-line (possibly a Masonic symbol) or a sun in splendour. These notes also juxtapose the motto that the society adopted – 'Nullius in verba', 'Nothing in words' or 'Nothing on authority' (a paraphrase of Horace), with such alternatives as the more mystical 'Ad Majorem Lumen', 'To the greater light', or 'Et Augebitur Scientia', 'And knowledge may be increased' (a quotation from Daniel 12:4 which had been the subject of millenarian speculation earlier in the century). Another possibility was 'Omnia probate', 'Try all things', which in fact echoes one of the mottos that Evelyn used on the bindings of his own books, 'Omnia explorate, meliora retinete', 'Explore everything, retain the best'. The alternatives that were evidently considered suggest a different, slightly more mystical, intellectual lineage for the Royal Society than that to which we are normally accustomed, though it is unclear whether Evelyn suggested all these ideas himself or (as is more likely) was merely recording deliberations in which he took part. Either way, this document illustrates his formative role in relation to the issues of symbolism and display that arose with the Sprat frontispiece.[7]

As for Evelyn's contribution to the society in its early years, perhaps most important was his *Sylva, or a Discourse of Forest-Trees* (1664), which has already been referred to because of John Beale's major contribution to it concerning cider-making. This was arguably the most successful publication produced under the society's auspices, which was issued in extended editions in 1670, 1679 and 1706. It represented a significant collaborative initiative on the part of the society's Fellows, in this case research carried out in response to a request from the Commissioners of the Navy for assistance in improving the country's timber supply, and Evelyn's task was to give an appropriate literary form to the disparate data on all aspects of the planting and nurture of trees that his colleagues collected, using what a contemporary described as his 'exquisite pen'.[8] The book drew to a significant extent on Evelyn's own horticultural and silvicultural experience in the 1650s, while at the end appeared an even more direct progeny of that, his *Kalendarium Hortense: or, the Gard'ners Almanac*, in fact an extract from his magnum opus on gardening, *Elysium Britannicum*, which he largely wrote at the end of the 1650s but which never got into print in his lifetime because of the distractions of his public responsibilities in the Restoration period.[9] It is also interesting that the society's minutes suggest that it was on topics of this kind that his colleagues saw him as especially expert, though there was obviously a miscellaneous streak in his contribution to the proceedings (also evidenced by his *Diary* entries on the meetings in question). If one looks at the specific tasks that he was asked to carry out, these disproportionately reflected his horticultural and related interests, while the principal paper that he was instrumental in presenting to the society from overseas also had an agricultural slant, namely the account of a Spanish machine for planting seed evenly which he transmitted to the society from the Earl of Sandwich in 1670 and which was published in *Philosophical Transactions*.[10]

In terms of Evelyn's own view of the society, an interesting document is provided by the defence of the society against its fashionable detractors that he inserted in the preface to the third edition of *Sylva* in 1679. His aim, since the society had been the 'chief *Promoter*' of that work, was 'to *vindicate* that *Assembly*, and consequently, the *Honour* of His *Majesty* and the *Nation*' by outlining the society's achievement in its first decade and a half of existence. Contrasting the society's output with the fruitless intellectual activity of its predecessors, he explained how, in 'as it were, *eviscerating* nature [and] disclosing the *resorts*, and springs of *Motion*', its Fellows had

> *collected* innumerable *Experiments, Histories* and *Discourses*; and brought in *Specimens* for the Improvement of *Astronomy, Geography, Navigation, Optics*; All the parts of *Agriculture*, the *Garden* and the *Forest*; *Anatomy* of *Plants, Mines* and *Ores*; *Measures* and *Æquations* of *Time* by accurate *Pendules*, and other motions, *Hydro-* and *Hygrostaticks*, divers *Engines*, Powers and *Automata*, with innumerable more *Luciferous* particulars, subservient to humane life.[11]

Evelyn's evaluation of the society and its achievement is revealing not least because of the extent to which he combines references to the fresh understanding of the natural world reached by such luminaries as Robert Hooke, Robert Boyle and Nehemiah Grew with an emphasis on the society's achievement in relation to agriculture, horticulture and other more practical pursuits. Also notable is Evelyn's emphasis on mensuration, and on machines and automata as a key part of the society's output, which is obviously relevant to the extent to which these appear in the frontispiece, as we shall see in due course.

Beyond that, Evelyn was an active and loyal member of the society, who was assiduous in his attendance at meetings for the remainder of his life, and who frequently served on the society's Council, from 1662 onwards. His prominent role is also reflected by the extent to which he was called on to become a member of various committees that the society set up in its early years. He was nominated to three of the eight committees covering all areas of its remit that the society set up in 1664, as referred to in the previous chapter, as also to others devoted to more specific topics such as 'equivocal generation' or the improvement of the English language.[12] He was also a member of the committee entrusted with supervising the society's 'repository', its collection of natural and artificial rarities, one of the most significant of the society's early institutional initiatives.[13] This had received a major boost early in 1666, when the society was recuperating in the aftermath of the Great Plague, by the purchase of a substantial collection of such rarities owned by the virtuoso and traveller, Robert Hubert or Forges, which had previously been on public display in London and the content of which had been itemised in a printed catalogue. Evelyn was also involved in a further, similar initiative in 1667, when it was he who persuaded Henry Howard, later 6th Duke of Norfolk, to present to the society the library of his grandfather, Thomas Howard, Earl of Arundel.[14]

Equally crucial, particularly by way of background to the frontispiece to Sprat's *History*, was Evelyn's participation in the society's programme for the 'history of trades', in other words, the project for recording and improving technological procedures which the society followed Bacon in seeing as central to its remit. Evelyn made a key contribution to this in 1661 by presenting to the society a 'Circle of *Mechanical* Trades', in which he listed and categorised a wide range of craft and industrial practices appropriate for study in this way, which derived from documents that he had drawn up in the previous decade.[15] His own offerings towards the society's enterprise included a couple of typical examples of such 'histories', namely his account of the making of marbled paper, and his 'Paneficium', a description of the methods of bread-making in France, while even more significant from our point of view was a further work which he contributed in 1662 and which was published in book form as *Sculptura: or the History, and Art of Chalcography and Engraving in Copper* (Fig. 3.3).[16]

In part, *Sculptura* was a pioneering contribution to the history of print-making that has recently received from Antony Griffiths the acclaim that it deserves, as almost the first attempt to provide information about artists and their works for the benefit of collectors.[17] But it also shows real sophistication in its account of artistic practice more generally, drawing on

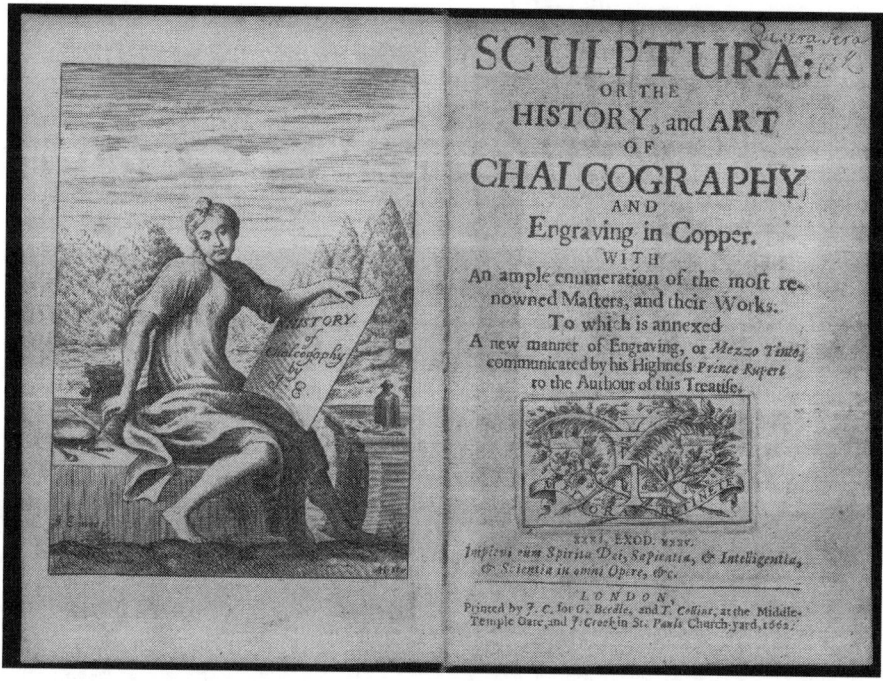

Figure 3.3 Evelyn's *Sculptura*, presented to the Royal Society in 1662 as part of its 'history of trades' programme. This copy from the collection of Sir Geoffrey Keynes has been inscribed with the motto, 'Que sera sera', 'Whatever will be, will be', by a former owner, Sir Thomas Knatchbull.

and seeking to propagate the ethos of Renaissance art which was so important to the practitioners and connoisseurs who were starting to form a native artistic tradition in England at this time. In addition, it was in this book that Evelyn published the first English account of mezzotint, a new print-making process that had been divulged to him by its English pioneer, Prince Rupert, who was evidently encouraged to provide Evelyn with details of the technique in connection with the history of trades programme.[18] This was a process which achieved a more painterly effect than traditional engravings by first mechanically 'grounding' the surface of the copper plate with a special tool so that the burrs on it would print an even black, and then progressively scraping and burnishing it to produce a design that is tonal rather than hatched as in an engraving. Evelyn's description in *Sculptura* in fact relates closely to some manuscript notes on the subject that he made in one of the commonplace books in which he recorded details of technical processes: he there illustrated the principal tools involved, one of which, the 'hatcher', he even applied to the surface of the page to show how the grounding effect was produced (Fig. 3.4).

Sculptura thus reflects not only Evelyn's commitment to the Royal Society but also his broader interest in the fine arts. This had earlier been much in

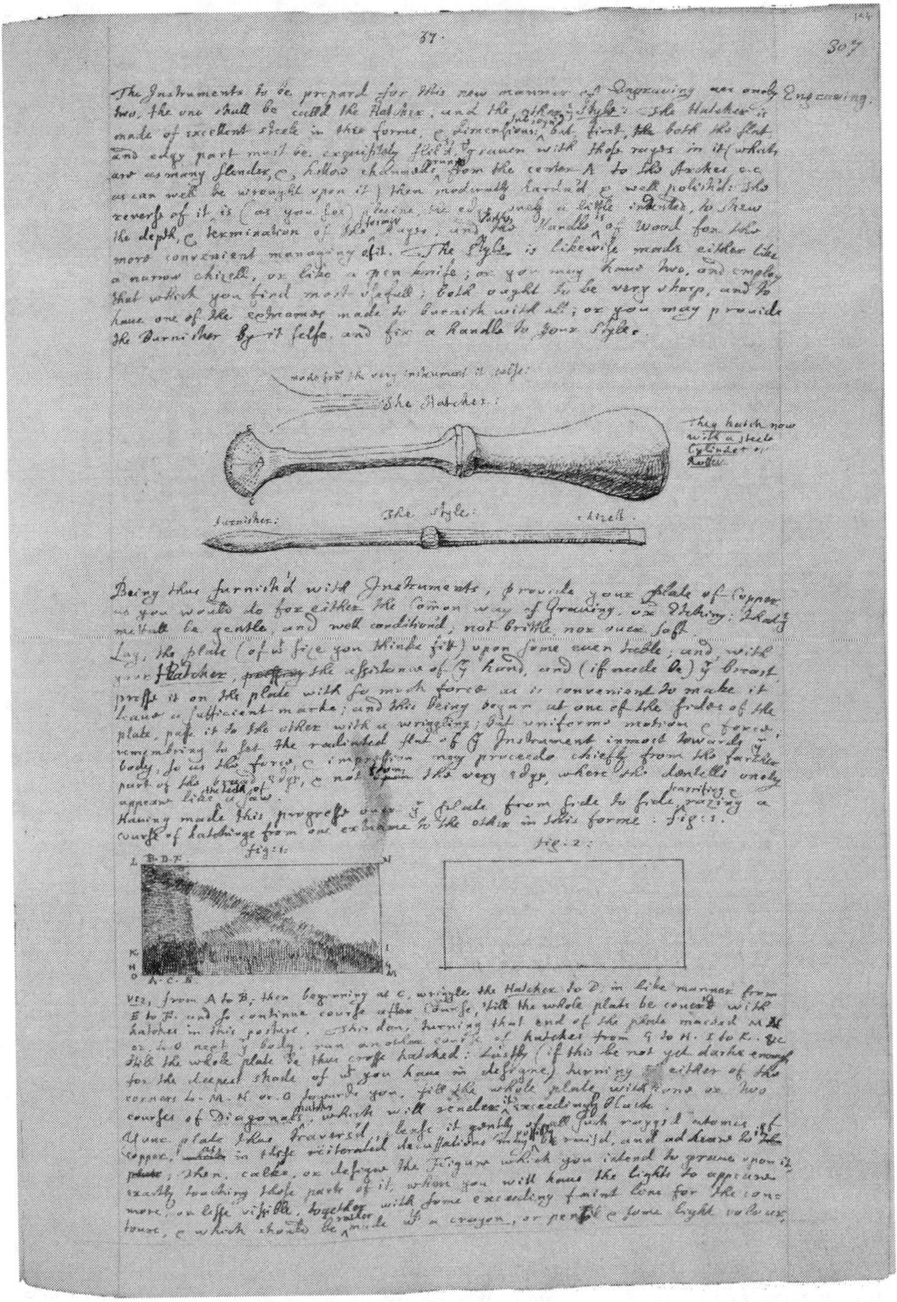

Figure 3.4 Evelyn's illustrated account of the process of mezzotint, including details of the tools and techniques involved.

evidence during his Grand Tour of Italy and other countries in the 1640s, as recorded in his *Diary*, when he patronised the young Carlo Maratti among others.[19] He had further honed his connoisseurial skills during the years that he spent in Paris around 1650, when he commissioned drawings of himself and members of his family by the distinguished French artist, Robert Nanteuil, who also produced a portrait print of Evelyn.[20] The project that Evelyn began with *Sculptura* was subsequently continued in two translations that he made of works by Roland Fréart, Sieur de Chambray – his *Parallel of the Antient Architecture with the Modern* (1664), which has already been mentioned, and his *An Idea of the Perfection of Painting*, which was to come out in 1668 and about which more will be said later – both of which sought to purvey the aesthetic ideas prevalent in contemporary France to an English audience.[21] All this, too, must have added to Evelyn's credentials in this regard for both Beale and his London colleagues, making him an appropriate designer for this crucial print.

Evelyn, Mary Evelyn and Wenceslaus Hollar

As has already been noted, in addition to writing about prints, Evelyn had in the past also made them, and two sets of etchings of landscapes by him survive, the first group imaginary, the second of views in Italy, both dating from 1649 (Fig. 3.5); in addition, in the early 1650s he produced a view of the Thames and another of his family home, Wotton, in Surrey.[22]

Figure 3.5 View of Naples from Mount Vesuvius, one of Evelyn's etchings of views between Rome and Naples, published in Paris in 1649.

On the other hand, thereafter his activity as an etcher ceased, and his activity in collecting prints seems also to have dwindled by the time he wrote *Sculptura*.[23] But he continued to execute drawings, and various sketches by him survive, though our knowledge of his achievement in this area is unfortunately restricted due to the dispersal of relevant material when the Evelyn family consigned his prints and drawings to Christie's in 1977 and they were scattered to the winds.[24] At least two drawings dated 1656 were included in the principal sale, one of them a charming but naïve sketch of his son, Richard, reading a natural history book, which reappeared at Christie's in 2012 and is now in the Cotsen Children's Library at Princeton University (Fig. 3.6).[25] In addition, both in the 1650s and later Evelyn did some quite competent, detailed drawings of equipment and the like to illustrate his contributions to the history of trades programme and his other treatises on technical subjects, including his incomplete *Elysium Britannicum*: an example is provided by his account of mezzotint which has just been described (Fig. 3.4).[26] Even if Evelyn's abilities in figure-drawing were limited compared with his skill in depicting landscapes and tools, one could nevertheless visualise him composing the design for the Sprat frontispiece.

Figure 3.6 Evelyn's drawing of his son, Richard, reading a natural history book, 1656.

Evelyn's wife, Mary, also had artistic skills. The 1977 sale included a number of drawings by her, one of them inscribed as executed at Paris in 1648.[27] Equally significant, in 1661 she made a copy of a drawing by Peter Oliver that Evelyn possessed and that also appeared in the 1977 sale (Fig. 3.14); as we will see, Evelyn notes in his *Diary* how he presented this copy to Charles II and how it had been 'wrought with extraordinary paines & Judgment' by his wife.[28] Even more relevant is the fact that in 1656 Mary designed the frontispiece for an earlier book by Evelyn, his English translation of the first book of Lucretius's *De Rerum Natura* (Fig. 3.7). This was etched by Wenceslaus Hollar, the experienced and prolific graphic artist who was to execute the Sprat print. Mary's design is in fact based rather closely on the title-frontispiece to a French translation of Lucretius that Michel de Marolles had published in Paris in 1650, albeit slightly softened and feminised.[29] On the other hand, commentators have noted the similarity of some of its details to the Sprat print, particularly in having a plinth flanked by two figures, above which a bust is being wreathed by angels (it has been claimed on the basis of a disparaging remark by Lucy Hutchinson that the bust was of Evelyn himself, though it was clearly intended to be that of Lucretius).[30] Indeed, evidently due

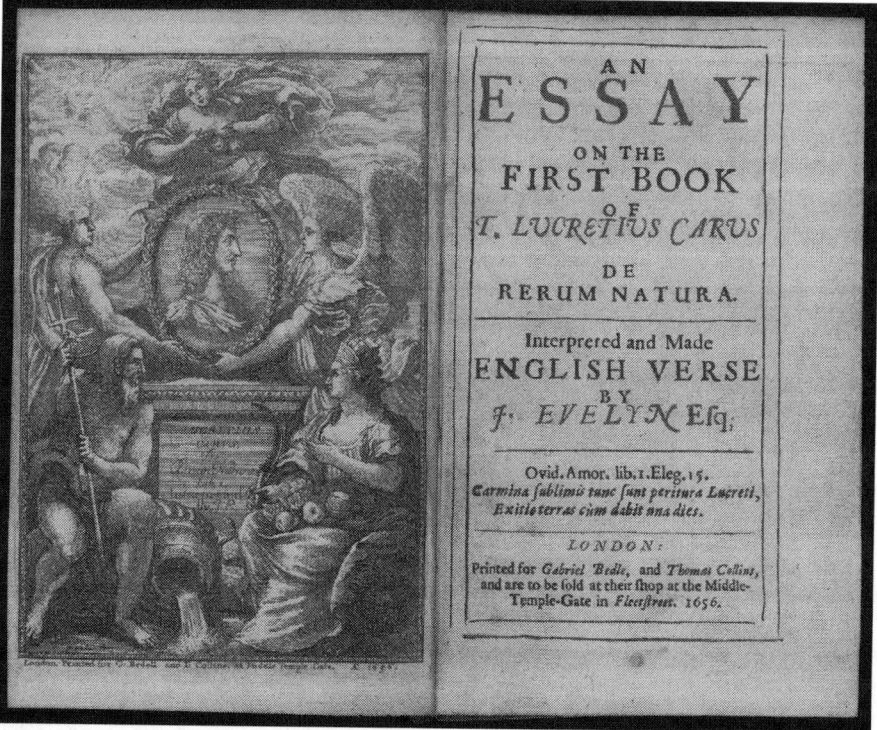

Figure 3.7 The title-page to Evelyn's translation of book 1 of Lucretius's *De Rerum Natura* (1656) with facing frontispiece designed by Mary Evelyn.

to this earlier initiative on Mary Evelyn's part, it has sometimes been stated that it was she rather than her husband who designed the Sprat frontispiece, though this is definitely not the case: the Lucretius plate is clearly signed 'Mary Evelyn inv.', in contrast to the 'Evelyn inv.' of the Sprat one, and there is no reason to doubt her husband's responsibility for the latter.[31]

Hollar's involvement in the earlier print would, however, have made him an obvious choice actually to etch the later one, and this was a decision in which Beale and the Royal Society would undoubtedly have concurred. As an acknowledged master of print-making, it would have seemed only appropriate for a specialist like Hollar to be enrolled to execute Evelyn's design, 'the professional presumably refining and improving the drawing of the amateur', as Richard Godfrey nicely put it of the Sprat frontispiece in his study of Hollar.[32] Here it is therefore appropriate briefly to introduce Hollar himself. Born in Prague and initially working in Germany, his English connection began through his patronage by Thomas Howard, Earl of Arundel, who employed Hollar in Cologne while on an embassy mission in 1636 and brought him to England in 1637. Hollar then became domiciled in England for the rest of his life, except for an interlude in the aftermath of the Civil War when he retreated to Antwerp (he also travelled to Tangier in 1669). Both before and after his sojourn in Antwerp, he produced an astonishing range of exquisite etched work that has often been seen visually to encapsulate the England of his time. His self-portrait (Fig. 3.8) shows him accompanied by his etching tools and it commemorates the period when he worked for the Earl of Arundel, since the print that he holds is a copy of a painting attributed to Raphael that was in Arundel's collection.[33]

Evelyn's own high view of Hollar's skills is borne out both by *Sculptura* and by an adulatory account of the artist that he included in his *Diary* when he wrote it up in his mature years. In addition, as early as 1644 he had been the subject of an elaborate dedication in a print made by Hollar, written by Hendrick van der Borcht the younger, a colleague of Hollar's who was also employed by Arundel and who corresponded with Evelyn (a portrait of Evelyn by him is currently in the National Portrait Gallery).[34] In 1645, while Evelyn was in Italy, van der Borcht had even suggested that he might supply subjects for the series of etchings of women from different parts of Europe on which Hollar was then engaged.[35] Much later, Hollar and Evelyn are known to have been in contact with one another in August 1667, when Hollar submitted a petition to Evelyn in connection with his plan for a gigantic map of London that never materialised. In it, he asked him for help in his approach to the government figures whose support he needed, making a rather desperate plea for help in which he referred to his 'unspeakable straites', and committed himself to Evelyn's 'worthy kindness, for the great Friendship you have of old testified to me, since the time I was in the service of that my most illustrious Patron and master Thomas Earle of Arrundel'.[36]

As the remark by Godfrey quoted above implies, Hollar's involvement further complicates the issue of responsibility for the design of the frontispiece, since it seems likely that Hollar must have had at least some input in its component

WENCESLAUS HOLLAR

*Gentilhomme ne a Prage lan 1607, a esté de nature fort inclin pʳ l'art de meniature principa-
lement pour esclaircir, mais beaucoup retardé par son pere, lan 1627, il est partij de Prage aijant
demeure en divers lieux en Allemaigne, il c est addonné pour peu de temps a esclaircir et apliquer
l eau forte, estant partij de Coloigne avec le Comte d'Arondel vers Vienne et dillec par Prage
vers l'Angleterre, ou aijant esté serviteur domestieque du Duc de Iorck, il s'est retire de la a cause
de la guerre a Anvers ou il reside encores.*
Io. Meyssens pinxit et excudit.

Figure 3.8 Self-portrait of Wenceslaus Hollar, with his etching tools and a plate after
a painting of St Catherine by Raphael owned by the Earl of Arundel but
now lost.

parts. For instance, he was almost certainly responsible for the sensitive portrait heads of Charles II, Bacon and Lord Brouncker, as we will see in the next chapter, and he must also have executed the detailed depictions of instruments that the plate includes. In fact, only a year previously, in 1666, Hollar had produced a striking portrait print of an instrument-maker, Elias Allen, with a variety of mathematical instruments in the background (Fig. 3.9).[37] On the

ELIAS ALLEN.

Apud Anglos Cantianus, iuxta **Cunnbridge** natus, Mathematicis Instrumentis ære incidendis sui temporis Artifex ingeniosissimus,

Obijt Londini, propè finem Mepsis Martij, Anno a Christo nato 1653 suæque ætatis

Figure 3.9 Hollar's portrait of the instrument-maker, Elias Allen, surrounded by instruments (1666).

other hand, it has been observed that many of Hollar's most successful prints were designed by others. As Richard Pennington, another commentator on his work, explains: 'Faced with a cathedral or a city, he could reproduce it perfectly as an artistic composition; but to create a unified composition out of his own mind was something he preferred not to do if he could help it.'[38]

In this connection, we need to interrogate the inscription on the frontispiece. This states clearly that 'Wenceslaus Hollar f[ecit] 1667', in other words that he actually etched the plate, and there is no reason to doubt that he did this in the year stated. As for 'Evelyn inv[enit]', this implies that Evelyn was responsible for 'conceiving' the overall design: Hollar used this term to describe works that others had devised, or that he had copied from earlier artists (for this, an alternative was 'pinxit', 'painted', as in the case of the Allen portrait, the (now lost) original of which was by van der Borcht). Sometimes, Hollar himself took credit for designing as well as etching prints, as with a series of images of sea storms that he produced in 1665, which bear the legend 'W. Hollar inv: et fec:'; another potential formula was 'delin[eavit]', 'drawn by', as with Hollar's views of Islington and other places on the outskirts of London dating from the same year, which bear the legend 'W. Hollar delin: et sculp:' – 'sculp[sit]', 'has engraved', being a synonym for 'fecit'.[39] The distinction between 'delineavit' and 'invenit' could imply the difference between a very exact original and a rougher sketch, and Evelyn's design for the Sprat frontispiece may perhaps have been lacking in detail even if clear in its overall conception, though, since it is lost, it is unfortunately impossible to be sure about this.[40] Lastly, the 'D.D.C.' that follows 'inv' in connection with Evelyn denotes a Latin dedication of the kind that is found in prints of the period: it is to be rendered 'donum [or 'dono'] dedit consecravitque', 'gives as a gift and consecrates', or some variant on that.[41]

It is also worth noting here the medium of the print, namely that it is an etching. Etching was a further graphic technique that differed from the more traditional process of engraving, as was the case with the novel process of mezzotint that Evelyn recorded in *Sculptura*. In the case of etching, the difference was that, rather than being directly incised on the copper plate by hand, the design was drawn with a needle through an acid-resistant ground (in this period, such an 'etching ground' usually consisted of a mixture of pine resin, asphaltum and beeswax melted together), after which the plate was corroded in either nitric acid or a mixture of various salts in strong vinegar. Hollar was a master of this art: in fact, the royalist traveller, Richard Symonds, took the trouble to record details of his etching technique, specifying his use of nitric acid.[42] As far as Evelyn is concerned, he explained the difference between etching and engraving in his *Sculptura*, and he might have provided further technical detail about the actual process in a second part of that work which would have been based on a pioneering treatise on the subject by the French artist, Abraham Bosse: but Evelyn never published this, deferring to the English translation of Bosse's work which was produced in 1662 by the famous engraver, William Faithorne the Elder, as *The Art of Graveing and Etching*.[43]

Here, it is important to note that it was still relatively novel in England at this time to use etching for book illustration: the first book to contain etched illustrations by an English hand had been Edward Benlowes's *Theophila* (1652), an example which was followed by Christopher Wren in the illustrations that he executed for Thomas Willis's *Cerebri Anatome*, published in 1664.[44] Indeed, Wren's 'very curious and exceeding speedy way of *Etching*' was referred to in the account of his various inventions that was included by Sprat in his *History*.[45] Etching was also used, mixed with engraving, in some of the plates for the celebrated *Micrographia* (1665) of Robert Hooke, Wren's former Oxford colleague and the Royal Society's first Curator of Experiments: this was perhaps the most famous illustrated book associated with the society in its early years, and it has long been celebrated for its stunning illustrations of a hidden world of tiny objects greatly magnified (perhaps most famously the flea).[46]

Evelyn's sources

Moving now to the design of the frontispiece, what can be said about this as a whole? As we have seen, some of its components may have been suggested by Beale, including the celebration of Bacon as the society's intellectual inspiration and of Charles II as its founder – though such concerns were hardly unique to him, reflecting as they did the ethos of the early Royal Society as a whole. Beyond this, in its use of perspective to place a central group of figures in an architectural background, the lineage of the print is to be found in Renaissance religious art, and perhaps particularly in depictions of the life of Christ and of the Holy Family. Evelyn himself had seen a range of works of art during his continental travels, as recorded in his *Diary*, while he had a large print collection, the remains of which were sadly dispersed along with his and his wife's drawings in the 1977 Christie's sales already referred to. Its exact content is unclear: though a summary listing survives as part of his extant library catalogue, the bulk of items that would fall into this category are subsumed under the general heading of 'Sacred history'.[47] However, he was certainly familiar with prints like those of the Passion or of the Life of the Virgin made by Albrecht Dürer, an artist for whose work he expressed great admiration in *Sculptura* – for instance, to take an example almost at random, the woodcut of the Death of the Virgin from his 'Life of the Virgin' of 1510 (Fig. 3.10). Like the Sprat frontispiece, this has a symmetrical architectural setting, grouped figures and perspectivised foreground, and, though this particular example lacks a landscape background and a chequered pavement, these are further features that often appear in images of this kind.[48] The overall effect of such echoes of Old Masters would have been to give the print a sense of sacred space, and this must surely have been Evelyn's intention.

A similar lineage can be traced for other components of the design, and not least the objects represented in its middle ground. These will be dealt with in detail in chapters 4 and 5, but here it is worth noting how the bookcase to the left of the central figures, though obviously a convenient device to enable books to be included, echoes comparable bookcases that were typical

Figure 3.10 Woodcut of the Death of the Virgin from Albrecht Dürer's 'Life of the Virgin' (1510).

in Renaissance depictions of St Jerome.[49] Moreover, pictures of scholars in their study – a common Renaissance genre – often included depictions of the tools of their trade, such as inkstands, penknives, spectacles and the like, while both these and pictures might be suspended from the walls, rather like the instruments in the Sprat frontispiece.[50]

Evelyn would also have been aware of the illustrations that had been produced during the previous century of instruments and technical processes, sometimes forming part of books, sometimes issued separately. He himself owned a copy of Hartman Schopper's classic illustrated account of crafts and occupations, *Panoplia: Omnium Illiberalium Mechanicarum aut Sedentariarum Artium Genera Continens*, published in Frankfurt in 1568, while in *Sculptura* he showed his awareness of perhaps the classic work of this genre, *Nova Reperta*, a series of engravings after designs by Johannes Stradanus

SCVLPTVRA IN ÆS.

Sculptor noua arte, bracteata in lamina   Sculpit figuras, atque prælis imprimit.

Figure 3.11 'Engraving in Copper'. Plate showing the interior of a plate-printing workshop, including two rolling presses, from *Nova Reperta*, published by Philips Galle after a design by Johannes Stradanus.

issued in the early 1590s; this celebrated discoveries of modern times, and one plate, 'Engraving in Copper', includes detailed depictions of various aspects of the print-making process (Fig. 3.11).[51] It is perhaps worth adding that at least one engraved title-page to a scientific text, J. M. Marci's *De Proportione Motus* (Prague, 1639), includes a range of devices and experiments slightly similar to those in the Sprat frontispiece, though it is unclear whether Evelyn knew it.[52]

Also worthy of note is Evelyn's evident fascination by actual assemblages of instruments of the kind that were depicted in works like these, as is revealed by such passages in his *Diary* as that when he saw the famous collection of the Jesuit scientist, Athanasius Kircher, at the Collegio Romano in 1644, a display that impressed many visitors to Rome.[53] Equally striking was his ecstatic reaction to the range of rarities at Wadham College, Oxford, that he was shown by John Wilkins, Warden of the college and their custodian, in 1654, including

> variety of *Shadows*, Dyals, Perspectives ... & many other artificial, mathematical, Magical curiosities: A Way-Wiser, a *Thermometer*; a monstrous *Magnes*, *Conic* & other *Sections*, a Balance on a demie Circle, most of them of his owne & that prodigious young Scholar, Mr. *Chr: Wren*.[54]

Returning to the frontispiece, when it comes to the central group of figures a much more precise source can be identified – indeed so precise that at first sight Evelyn's design seems almost like an act of plagiarism. The work in question is a component of a book that Evelyn is known to have owned, namely Nicolas Chaperon's set of etchings of the biblical scenes that Raphael had depicted in the loggia at the Vatican in the second decade of the sixteenth century.[55] Chaperon was a member of a colony of French artists who worked in Rome in the mid-seventeenth century which also included Nicolas Poussin; his celebratory volume concerning the Vatican loggia appeared in 1649, entitled *Sacrae Historiae Acta a Raphael Urbin in Vaticanis xystis ad Picturae Miraculum Expressa*. Mainly, this comprises a series of fifty-two etchings of the various biblical scenes with which Raphael had decorated the loggia, most of them from the Old Testament though with a few from the New (Fig. 3.12a and Fig. 3.12b); but at the start was an elaborate title-page and frontispiece, and it was in the latter that Evelyn found the model that he wanted.

This is reproduced here as Fig. 3.13, and, as will be seen, we have a composition which replicates many facets of the Sprat frontispiece very closely indeed. At the centre is a round column or drum set on a square plinth with mouldings at top and bottom and with an inscription in capitals to which a figure is gesticulating with his right hand. The figure in question is dressed in a sort of toga, but his left, gartered leg protrudes from under it and both his shoes are visible. On top of the column is a bust on a pedestal, which is being crowned with a wreath by a winged figure representing Fame, dressed in flowing robes but with her left arm bare, and with a trumpet held in her right hand which rests on her shoulder. Less of this figure is visible on Evelyn's print than on Chaperon's because an extra figure has been introduced in front of it in the form of the seated figure of Lord Bacon, while other details also differ, as we will see when we turn to consider Evelyn's additions and alterations. But the overall similarity of the basic composition is striking, and it is perhaps worth noting in passing a further overlap in the use of an architectural surround, though with different details, and even in the way in which a landscape vignette appears behind the angel to the right – though it is differently placed, while the scene in Evelyn's print contrasts with Chaperon's depiction of St Peter's and its Roman surroundings.

Before going on to Evelyn's adaptation of it, let us consider the components of Chaperon's plate and their significance in their own right. In this case, the bust is of Raphael, and it is based on various portraits of him derived from that in the Villa Lante at Rome, including a print by Giulio Bonasone and the famous Double Portrait of Raphael and a friend now in the Louvre.[56] The inscription on the column commemorates both Raphael himself (echoing his epitaph at the Pantheon in Rome) and also Chaperon's role in perpetuating his fame; this is echoed by the inscription at the bottom of the design.[57] The figure to the left of the column is a self-portrait of

Figure 3.12a and 3.12b Two of the etchings of Raphael's loggia at the Vatican that make up the bulk of Chaperon's *Sacrae Historiae Acta a Raphael Urbin in Vaticanis xystis ad Picturae Miraculum Expressa* (1649), showing God introducing Adam to Eve (no. 5) and the building of Solomon's temple (no. 48).

Figure 3.13 The frontispiece to Chaperon's *Sacrae Historiae*.

Chaperon himself, though the way his head is shown is heavily influenced by self-portraits of his mentor, Poussin, while the flowing robes that he wears are influenced by Michelangelo's depiction of Moses in his tomb for Pope Julius II in the church of San Pietro in Vincoli in Rome (this is also echoed by the figure of Isaiah on the adjacent title-page to Chaperon's book).[58] As for the figure of Fame, this evokes the winged putti bearing wreaths that appear in two other works by Poussin, his 'Inspiration of the Lyric Poet' at Hanover and his 'Inspiration of the Epic Poet' in the Louvre.[59] On the other hand, Chaperon has adapted it by denoting Fame as a more mature figure, including such traditional iconographic trimmings as a trumpet, and here his design appears to owe more of a debt to a title-page designed by Rubens and published in 1617: this is similarly focussed on a moulded stone drum on which an inscription appears and on which a figure is being crowned with a wreath by a comparable figure of Fame wearing flowing robes – albeit it is a figure rather than a bust that is being wreathed, while Fame bears a quill rather than a trumpet.[60] In the Chaperon print, an artist's pallet and brushes lie on the ground beside him, while a board propped up against the architectural surround shows an emblematic device with a motto, evidently associated with his patron, the French politician, Gilles Renard, to whom the volume was dedicated.[61]

The role of Raphael

It is an effective composition, and Evelyn must clearly have felt that here he had found an overall scheme that would suit his purposes; indeed, this might have been a sufficient reason in itself for his adaptation of it. But can any conclusions be drawn from the nature of his source, and in particular might he have been making a conscious statement by following Chaperon's design so closely, since this was something that contemporary connoisseurs would surely have noticed?[62] Raphael – whose bust is being wreathed at the centre – was, of course, frequently acclaimed as the greatest painter of his age. He was also responsible for one of the greatest 'philosophical' paintings of all time in the form of 'The School of Athens' at the Vatican, with its heroic depiction of Plato, Aristotle and other ancient interpreters of nature. Evelyn had been predictably ecstatic about various paintings by Raphael that he had seen while on his continental travels in the 1640s, using words to describe them like 'incomparable', 'very rare', 'inimitable' or 'celebrated', and singling out one of them as 'a piece which all the world admires'.[63] In this, he followed his mentor, Thomas Howard, Earl of Arundel, who drew attention to various works by Raphael in his travel notes for Evelyn and one of whose most prized possessions was a portrait that he believed to be by Raphael.[64]

A similarly high view of Raphael and of the engravings of his work by Marcantonio Raimondi appears in Evelyn's *Sculptura*, while Raphael was also seen as outstanding in the book on modern art by Roland Fréart, Sieur de Chambray, that Evelyn translated into English in 1668 as *An Idea of the Perfection of Painting*. Fréart's book aspired to the revival of the principles of classical art, claiming that it was Raphael who had come closest to this in modern times, and his book was focussed on a discussion of a series of works by Raphael, including 'The School of Athens'. In Fréart's words, Raphael was 'the most excellent of the *Modern* Painters, and universally so reputed by those of the *Profession*'; for this reason, he used Raphael's work as 'so many *Demonstrations* of the absolute necessity of exactly observing the *Principles* which we have establish'd in this *Treatise*'.[65]

It is therefore perhaps not surprising that it was of a picture attributed to Raphael that Evelyn presented Charles II with a miniature version executed by his wife, Mary, in 1661, noting in his *Diary* how the king was 'infinitely pleas'd with it, & caus'd it to be placed in his Cabinet amongst his best Limmings'.[66] This was a copy of an artwork owned by Evelyn which is interesting in itself and which further illustrates his enthusiasm for Raphael. The item in question was an ink and grey wash copy made in 1631 by the miniature painter, Peter Oliver (son of the celebrated miniaturist, Isaac Oliver), of a painting in the collection of Charles I of the Madonna and Child with the infant St John then thought to be by Raphael. Oliver made a series of such accomplished copies of Old Masters owned by the king from 1628 onwards, which Charles himself valued highly; a number are in the Royal Collection today.[67] Tellingly, Evelyn had spent £20 on purchasing Oliver's 'Raphael' copy, and, remarkably, it is still extant and is reproduced here (Fig. 3.14).[68] By this time the original was no longer in the country, having been acquired in the dispersal of Charles's collection on behalf of Philip IV of Spain. It remained in Spain till the Napoleonic period, when it was purchased by the English collector, William Bankes, and today is back in England again, at Kingston Lacy in Dorset (Fig. 3.15). It is now thought to be by a follower of Raphael rather than by Raphael himself.[69]

This episode is crucial because it may be seen in the context of Evelyn's attempts to encourage the king to patronise the fine arts in England and thus to emulate the cultural example of France and Italy, to which Charles only fitfully responded.[70] Evelyn's translation of Fréart's *Parallel of the Antient Architecture with the Modern* (1664) was dedicated to Charles, and he there noted how 'as in all other Princely and magnificent things your *Notices* are extraordinary, so I cannot but augure of their *effects*', claiming that architecture 'ows her *renascency* amongst *Us* to your *Majesties* encouragements, and to as many of those *Illustrious* Persons as by their large and magnificent *Structures* transcribe your Royal *Example*'.[71] Later, Evelyn saw *An Idea of the Perfection of Painting* as having a complementary aim, in that it 'does, I think, perfectly consummate that *designe* of mine, of recommending to our *Countrey*, and especially to the *Nobless*, those *Three* Illustrious and magnificent *Arts*, which are so *dependent* upon each other', namely architecture, sculpture and painting.[72] All of this illustrates Evelyn's sense of mission to encourage his fellow countrymen, from the monarch downwards, to emulate

Figure 3.14 Peter Oliver's ink and grey wash copy, made in 1631, of the painting in the collection of Charles I of *The Holy Family with the Infant St John in a Landscape*, believed to be by Raphael (the original is 241 by 174 mm). It was of this drawing, later owned by Evelyn, that a copy in miniature made by his wife, Mary, was presented to Charles II in 1661. Mary's copy is now lost, but it may have resembled the 'Raphael' original (see Fig. 3.15) in being coloured and framed.

Figure 3.15 The Holy Family with the Infant St John in a Landscape, formerly attributed to Raphael, now to his circle. Once in the collection of Charles I,
this painting is now at Kingston Lacy, Dorset, in an elaborate frame
celebrating its former owners commissioned by William Bankes.

the best of continental culture – something that was arguably just beginning
to come about through the great architectural enterprise on which Evelyn's
fellow virtuoso, Christopher Wren, was at this time embarking.[73] Indeed,
from this point of view a telling comment on Evelyn was made in 1668 by a
representative of the cultural milieu to which he aspired, the Italian traveller,

Lorenzo Magalotti, who slightly disdainfully described Evelyn as 'a great expert on agriculture and in the judgment of the English equally esteemed as a connoisseur of painting and architecture'.[74]

In this connection, it is appropriate to return to the Royal Society and to Evelyn's *Sculptura*. We have already seen how this was intended as part of the society's 'history of trades' programme, and this is significant here because Evelyn clearly believed that the improvement of knowledge to which the society aspired should include not only the understanding of nature but also the pursuit of the fine arts.[75] We should beware of a modern sense of the separation between science and art obscuring the unity of vision of a man like Evelyn. To him, Raphael would have seemed as proper an intellectual hero for the Royal Society as Copernicus, and this may be one of the messages that the frontispiece is trying to convey. Indeed, the implication may even be that Evelyn dreamt of making the Royal Society the English equivalent not only of the Académie des Sciences that Jean-Baptiste Colbert inaugurated in France in 1666 but also of the Académie de Peinture et de Sculpture, which went back to 1648 although it had been revitalised under Colbert's influence in 1663.[76]

Comparable sentiments are to be found among others associated with the society at this time. When Henry Oldenburg reviewed Evelyn's *Idea of the Perfection of Painting* in *Philosophical Transactions* in 1668, he expressed the hope that the book would 'animate many among us to acquire a perfection in Pictures, Draughts and Chalcography, equal to our growth in all sorts of Optical Aydes, and to the fulness of our modern Discoveries'.[77] It is also significant that in 1666 the society had set up a committee to consider papers concerning pigments and other aspects of artists' techniques which had been collected by the physician and connoisseur, Sir Theodore de Mayerne, earlier in the century and were divulged to the society by his godson, the medical Fellow, Sir Theodore de Vaux, while later the same year a committee including Evelyn was tasked with investigating aspects of painting on the basis of inquiries made by another Fellow who was also an artistic connoisseur, Thomas Povey.[78] In connection with this, Povey presented a paper to the society in which he wrote: 'Nor is the generous subject of painting unworthy of some part of your care and research, which hath been the study, delight and ornament of all ages and nations, where peace and civility have not been abandoned.' In his view, this deserved 'an entire history, to be conducted by such an influence as yours', which would show that the society was concerned not 'with curiosity and speculation only', but would instead 'leave something new or improved to the succeeding world'.[79]

In addition, in a passage in his *History*, Thomas Sprat expatiated on the appropriateness of the Royal Society being invited to hold its meetings at Arundel House in the aftermath of the Fire of London, when Gresham College became unavailable because it was taken over by the City. Arundel House was where what was left of the collections of art and sculpture of Evelyn's mentor, the great Earl of Arundel, were preserved, and of its current owner, Henry Howard, Duke of Norfolk, Sprat wrote:

> by entertaining these new discoveries under his Roof, his *Family* deserves the double praise of having cherish'd both the old, and new *Learning*; so that now methinks in *Arundel* house, there is a perfect representation, what the Real *Philosophy* ought to be: As there we behold new *Inventions* to flourish amongst the *Marbles*, and *Images* of the *Dead*: So the present *Arts*, that are now rising, should not aim at the destruction of those that are past, but be content to thrive in their company.[80]

Sprat's views were echoed by Evelyn himself when dedicating his *Idea of the Perfection of Painting* to the Duke of Norfolk a year later.[81]

Conclusion

As he transformed Raphael into Charles II and Chaperon into the Royal Society's first President, Evelyn may thus consciously have been transmuting the ambitions of Renaissance art into those of the new science. But of course he also made changes, most notably the insertion of a second figure flanking the column to balance that of Chaperon/Brouncker, in the form of the seated image of Lord Bacon attired in his robes as Lord Chancellor and holding his purse of office. Obviously, this significantly enhanced the print's relevance to the Royal Society at the expense of its Renaissance antecedents, and it would have been especially appropriate to a composition intended for John Beale's project, devoted as it was to 'Lord Bacon's Elogyes'. In addition, the inscriptions were naturally changed to identify the new protagonists, that on the column alluding to Charles as 'Author & Patron' of the society, while the inscriptions that were added on the chequered pavement in the foreground identified Brouncker as its President and Bacon as 'Renewer of the Arts', the progenitor of the intellectual revolution which the society saw itself as putting into effect. It was obviously equally appropriate for Evelyn to include the Royal Society's coat of arms in place of the emblem of Chaperon's patron.

Lastly, the overall balance of the composition was altered by bringing the column down within it so that its top, and the base of the bust, are shown in perspective rather than in profile. This meant that the strong thrust from left to right of Chaperon's design was lost and the print was instead made more symmetrical, with a larger proportion of it appearing above and to the right of the central group; at the same time, the architectural surrounding was altered and elaborated. On the other hand, this provided an opportunity for ingenious spatial design which is almost entirely unprecedented in Chaperon's print. In particular, the wreath above Charles II's head is at the centre of a vortex of triangulation which emphasises his significance, while the composition is further articulated by diagonal lines: a strong axis opens up from the central point above the wreath, running down the telescope, wall and bookcase on the left, while a matching, if less marked, axis stretches down Bacon's profile to the right. The triangular shape thus formed is echoed by the diagonals emanating from the tiles of the floor in the foreground, and the whole image is full of triangulations which intersect with one another and make it aesthetically pleasing.[82] The careful geometrical planning of the frontispiece thus enhances the spatial interest of the composition, aligning it with the analogues in Renaissance sacred art that have already been referred to. In addition, Evelyn provided himself with the setting for a host of significant details, and to these we must now turn.

Notes

1 Birch, *Royal Society*, vol. 1, pp. 4, 8.
2 Hunter, *Establishing*, pp. 16–17.
3 Evelyn, *Diary*, vol. 3, p. 306.
4 John Evelyn (trans), Gabriel Naudé, *Instructions concerning Erecting of a Library* (London, 1661), sigs. A2–7.
5 Beale to Evelyn, 28 April, 24 May 1666, Add. MS 78312, fols. 36–7, 38–9 (quotation at fol. 37).
6 Add. MS 78344, fol. 114.
7 Hunter, *Establishing*, pp. 17, 41–2. On the actual motto adopted, see also https://blogs.otago.ac.nz/emxphi/2012/02/nullius-verba-nihil-verbis-sapere-aude/.
8 Quoted in Michael Hunter, *Science and Society in Restoration England* (Cambridge, 1981), p. 93. See also pp. 100–101. For an account of the book see William T. Lynch, *Solomon's Child: Method in the Early Royal Society* (Stanford, 2001), ch. 2. See also Birch, *Royal Society*, vol. 1, pp. 13, 111, 114, 117–18, 120, 179, 347.
9 See Frances Harris, 'The Manuscripts of the "Elysium Britannicum"', in John E. Ingram (ed.), *John Evelyn, Elysium Britannicum, or The Royal Gardens* (Philadelphia, 2001), pp. 13–19, on pp. 15–16.
10 'A Letter . . . Concerning the Spanish *Sembrador*', *Phil. Trans.*, 5 (1670), 1055–65, and Birch, *Royal Society*, vol. 2, p. 425. Cf. Birch, vol. 1, pp. 16, 25, 53, 78, 111, 177, 316; vol. 2, pp. 73, 180, 298–9, 324, 374, 418–19, 440, 468. See also Evelyn, *Diary*, vol. 3, passim.
11 Evelyn, *Sylva* (3rd edition, London, 1679), 'To the Reader', sigs. A1, A3. He ended the sentence by citing Glanvill's *Plus Ultra* (London, 1668). For a similar commentary from the period when Sprat's *History* was being prepared, see Evelyn to Abraham Cowley, 12 March 1667, in Douglas D. C. Chambers and David Galbraith (eds), *The Letterbooks of John Evelyn* (2 vols., Toronto, 2014), vol. 1, pp. 435–6.
12 Birch, *Royal Society*, vol. 1, pp. 212 (and pp. 213, 217, 238, 448 for his attendance), 406–7, 500. For other committees see vol. 1, pp. 15, 207, 327; vol. 2, pp. 29, 230, 371. See also Hunter, *Establishing*, ch. 3.
13 Birch, *Royal Society*, vol. 2, p. 73. See Michael Hunter, 'Between Cabinet of Curiosities and Research Collection: The History of the Royal Society's "Repository"', in Hunter, *Establishing*, ch. 4.
14 See Linda Levy Peck, 'Uncovering the Arundel Library at the Royal Society: Changing Meanings of Science and the Fate of the Norfolk Donation', *Notes & Records*, 52 (1998), 3–24, esp. p. 4.
15 Birch, *Royal Society*, vol. 1, p. 12. Cf. Michael Hunter, 'John Evelyn in the 1650s: A Virtuoso in Quest of a Role', in Hunter, *Science and the Shape of Orthodoxy:*

Intellectual Change in Late Seventeenth-Century Britain (Woodbridge, 1995), pp. 67–98, on pp. 75–6. See also Gillian Darley, *John Evelyn: Living for Ingenuity* (New Haven and London, 2006), esp. ch. 8. On the Royal Society's history of trades programme see Hunter, *Science and Society*, ch. 4, and K. H. Ochs, 'The Royal Society of London's History of Trades Programme: An Early Episode in Applied Science', *Notes & Records*, 39 (1985), 129–58.

16 Birch, *Royal Society*, vol. 1, pp. 33, 69, 83, 86; vol. 2, p. 19. For 'Paneficium' see Hunter, *Science and Society*, pp. 102–3; for the account of marbled paper, Hunter, 'John Evelyn in the 1650s', p. 78.

17 Antony Griffiths, 'John Evelyn and the Print', in Frances Harris and Michael Hunter (eds), *John Evelyn and His Milieu* (London, 2003), pp. 95–113; C. F. Bell (ed.), *Evelyn's Sculptura. With the Unpublished Second Part* (Oxford, 1906), passim. See also Craig A. Hanson, *The English Virtuoso: Art, Medicine and Antiquarianism in the Age of Empiricism* (Chicago, 2009), pp. 80–3 and ch. 2, passim.

18 See the full account in Ben Thomas, 'Noble or Commercial? The Early History of Mezzotint in Britain', in Michael Hunter (ed.), *Printed Images in Early Modern Britain: Essays in Interpretation* (Farnham, 2010), pp. 279–96. For a transcript of the text, which is now Add. MS 78340, fol. 154, see Orovida C. Pissarro, 'Prince Rupert and the Invention of Mezzotint', *Walpole Society*, 36 (1956–8), 1–9, on pp. 3–5. On mezzotint more generally, see Antony Griffiths, *Prints and Printmaking: An Introduction to the History and Techniques* (2nd edition, London, 1996), pp. 83ff., and Ad Stijnman, *Engraving and Etching 1400–2000: A History of the Development of Manual Intaglio Printmaking Processes* (London, 2012), pp. 184ff.

19 Evelyn, *Diary*, vol. 2, passim. For his patronage of Maratti, see esp. vol. 2, pp. 223, 247, and Griffiths, 'John Evelyn and the Print', p. 96.

20 On Evelyn's relations with Nanteuil, see Audrey Adamczak, *Robert Nanteuil c. 1623–78* (Paris, 2011), pp. 38ff., 134ff., 300ff. The pencil portrait of Evelyn himself is reproduced here as Fig. 3.1; the other Nanteuil drawings have most recently been reproduced in Chambers and Galbraith, *Letterbooks of John Evelyn*, vol. 1, pp. 94–7; the print appears as the frontispiece (it is also reproduced in Antony Griffiths, *The Print in Stuart Britain, 1603–89* (London, 1998), p. 132 (catalogue entry 81), and in Adamczak, *Robert Nanteuil*, p. 41; she reproduces Nanteuil's drawings on pp. 40 and 135–6).

21 On Evelyn's translations of Fréart see Hanson, *English Virtuoso*, pp. 83–90, and Ben Thomas, 'John Evelyn's Project of Translation', *Art in Print*, 2, no. 4 (November–December 2012), 28–34.

22 See Antony Griffiths, 'The Etchings of John Evelyn', in David Howarth (ed.), *Art and Patronage in the Stuart Courts: Essays in Honour of Sir Oliver Millar* (Cambridge, 1993), pp. 51–67. For related material unknown to Griffiths, see ch. 6, n. 23 in this book. For watercolour views of Wotton House by Evelyn, see Chambers and Galbraith, *Letterbooks of John Evelyn*, vol. 1, plates 1–2.

23 Griffiths, 'John Evelyn and the Print', pp. 105–6 and passim. What was by then left of Evelyn's print collection was sold at Christie's on 29 June 1977 (lots 1–126) and 26 July 1977 (lots 156–99).

24 See Christie's catalogue, *Important English Drawings, Watercolours and Pastels*, 14 June 1977, lots 152–68. Lot 152 was the Peter Oliver drawing of the Madonna and Child with the infant St John reproduced here as Fig. 3.14. There were also drawings by Francis Barlow (lots 159–60), William Faithorne (lots 162–3), Charles Beale (lot 164) and others. Other drawings owned by Evelyn, including the one of the Arch of Titus by Carlo Maratti (on which see Griffiths, 'John Evelyn and the Print', p. 96), appeared in sales on 6 July 1977 (lots 1–20) and 12 July 1977 (lots 48–51).

25 See Christie's, South Kensington, catalogue, *Travel, Science and Natural History*, 25 April 2012 (sale 4826), lot 15. This was lot 153 in the 14 June 1977 sale. The other item dated 1656 is a landscape after Titian, lot 154 in the 14 June 1977 sale: this is now at the Yale Center for British Art: see Richard T. Godfrey, *Wenceslaus Hollar: A Bohemian Artist in England* (New Haven and London, 1994), p. 149, where it is reproduced. For other figure-drawings by Evelyn, see lots 155–6.

26 See Add. MS 78340, fol. 154; for other examples see Hunter, 'John Evelyn in the 1650s', p. 77. Fig. 5.26 in this book; and Evelyn, *Elysium Britannicum*, passim.

27 Christie's, 14 June 1977, lots 157, 167–8. Lot 158 comprised drawings, one dated 1687, by Evelyn's daughter, Susannah, on whom see Carol Gibson-Wood, 'Susannah and Her Elders: John Evelyn's Artistic Daughter', in Harris and Hunter (eds), *John Evelyn and His Milieu*, pp. 233–54 (though she was unfortunately unaware of these drawings).

28 Christie's, 14 June 1977, lot 152. Evelyn, *Diary*, vol. 3, p. 287. See p. 47 in this book.

29 For the Marolles title-frontispiece and Mary Evelyn's adaptation of it see Cosmo A. Gordon, *A Bibliography of Lucretius* (London, 1962; reissued Winchester, 1985), pp. 154, 174 and Plate 16. For the Mary Evelyn print, see Keynes, *John Evelyn*, pp. 43–4; Pennington, *Hollar*, no. 2677; *New Hollstein*, no. 1435; A. F. Johnson, *Catalogue of Engraved and Etched English Title-Pages* (London, 1934), p. 30 (Hollar, no. 22). See also Frances Harris, 'Living in the Neighbourhood of Science: Mary Evelyn, Margaret Cavendish and the Greshamites', in Lynette Hunter and Sarah Hutton (eds), *Women, Science and Medicine 1500–1700* (Stroud, 1997), pp. 198–217, on p. 201 (the source of the observation that elements in the image had 'undergone a significant softening and feminisation').

30 For its similarities to the Sprat frontispiece see H. C. Levis, *Extracts from the Diaries and Correspondence of John Evelyn and Samuel Pepys Relating to Engraving* (London, 1915), pp. 140–1, and Katherine S. van Eerde, *Wenceslaus Hollar: Delineator of His Time* (Charlottesville, 1970), p. 49. For the suggestion that the head is that of Evelyn, see Keynes, *John Evelyn*, p. 43, and de Beer's note in *Diary*, vol. 3, p. 173, both citing Sir Charles Firth's edition of Lucy Hutchinson's *Memoirs*: her comment appears in the dedicatory epistle to her own translation of Lucretius, Hugh de Quehen (ed.), *Lucy Hutchinson's Translation of Lucretius, 'De Rerum Natura'* (London, 1996), p. 23.

31 For the erroneous suggestion that the Sprat frontispiece was designed by Mary Evelyn, see Nellie B. Eales, *The Cole Library of Early Medicine and Zoology* (2 vols., Reading, 1969–75), vol. 1, p. 112 (no. 634), and Christie's catalogue, *The Evelyn Library, Part III: M–S*, 15–16 March 1978, lot 1405, relating to Evelyn's copy of Sprat, which was sadly stolen: see ch. 6, n. 10 in this book. Uncertainty on the matter is registered by Levis, *Extracts*, p. 141.

32 Godfrey, *Wenceslaus Hollar*, p. 22.

33 Pennington, *Hollar*, no. 1419, *New Hollstein*, no. 1058. It was published in 1649 as part of Jan Meyssens's *Image de Divers Hommes*. The city in the background has sometimes been thought to be Prague, his birthplace, but this seems unlikely. For secondary literature on Hollar, see ch. 1, n. 1 in this book.

34 Bell (ed.), *Evelyn's Sculptura*, pp. 81–2; Evelyn, *Diary*, vol. 1, pp. 21–2; Pennington, *Hollar*, no. 1393; *New Hollstein*, no. 563; Robert Harding, 'John Evelyn, Hendrick van der Borcht the Younger and Wenceslaus Hollar', *Apollo*, vol. 144, no. 2 (issue 414, August 1966), 39–44. For the van der Borcht portrait see Chambers and Galbraith, *Letterbooks of John Evelyn*, vol. 1, plate 3.

35 Harding, 'John Evelyn', pp. 41, 42.

36 Ibid., pp. 41–2, 43–4. The document is in a scribal hand (see p. 44 n. 15). For the map see Simon Turner, 'Hollar's Prospects and Maps of London', in Hunter, *Printed Images in Early Modern Britain*, pp. 145–66, on pp. 151–3.

37 For the print of Allen, see Pennington, *Hollar*, no. 1338; *New Hollstein*, no. 1928: the instruments included are mostly those appropriate to a mathematical instrument maker, whereas those in the Sprat frontispiece are appropriate to a research community. See also Hester Higton, 'Portrait of an Instrument-Maker: Wenceslaus Hollar's Engraving of Elias Allen', *British Journal for the History of Science*, 37 (2004), 147–66.

38 Pennington, *Hollar*, p. xliv. It is perhaps worth noting that we have looked at prints by Hollar dating from the years preceding the Sprat frontispiece in search of analogous features, but only rather general similarities have come to light. For instance, in Hollar's illustrations to Sir Robert Staplyton's translation of Juvenal, *Mores Hominum* (London, 1660), or *The Fables of Æsop* (London, 1665), or *Æsopic's* (London, 1668), Pennington, *Hollar*, nos. 333–90, 391–408, 429–45, *New Hollstein*, nos. 1752–68, 1857–1913, 2005–34, such features as chequered floors, architectural surrounds and gesticulating figures appear, but these seem only to reflect the legacy of Renaissance art that would have been equally familiar to Evelyn, as suggested later in this chapter.

39 Pennington, *Hollar*, nos. 915–20, 1273–6, *New Hollstein*, nos. 1827–32, 1839–42 (in some of the sea storms as well as the views, this is combined with 'sculp' rather than 'fecit'). See also the glossary in Stijnman, *Engraving and Etching*, pp. 415–18.

40 For a surviving drawing for a print of the period which is very similar to the print as executed, see n. 60 in this chapter.

41 See C. T. Lewis and Charles Short, *A Latin Dictionary* (Oxford, 1879, repr. 1927), p. 509, col. 2. We are grateful to Antony Griffiths for his advice on this matter.

42 Symonds's comments are quoted from the version of them preserved by George Vertue in various of the studies referred to in ch. 1, n. 1, e.g., Pennington, *Hollar*, p. xlix, or Godfrey, *Wenceslaus Hollar*, pp. 18–19 (but see Godfrey, p. 33 n. 65, concerning an error by Pennington): *Walpole Society*, 18 (1930) (*Vertue Note Books*, vol. 1), p. 112. For a full transcription of the notes by Symonds of which they form part from the original manuscript, British Library Egerton 1636, see Mary Beal, *A Study of Richard Symonds, His Italian Notebooks and Their Relevance to Seventeenth-Century Painting Techniques* (New York, 1984), pp. 211ff., with a commentary on his notes on etching, including those derived from Hollar, on pp. 203ff. In his *A Description of the Works of Wenceslaus Hollar* (2nd edition, London, 1759), pp. 133ff., Vertue also included a further document associated with Hollar concerning etching technique, though the authenticity of this is uncertain: see Pennington, *Hollar*, p. lxi.

43 Bell (ed.), *Evelyn's Sculptura*, pp. 9–10 and Part 2, passim, the introduction to which explains about Evelyn's suppression of the second part of the work in deference to Faithorne and the rediscovery and publication of Evelyn's manuscript.

44 See Nathan Flis, 'Drawing, Etching and Experiment in Christopher Wren's Figure of the Brain', *Interdisciplinary Science Reviews*, 37 (2012), 145–60 (though he is wrong to assert on p. 145 that *Cerebri Anatome* was published by the Royal Society). On the techniques involved see also Griffiths, *Prints and Printmaking*, pp. 56ff., and Stijnman, *Engraving and Etching*, pp. 45ff.

45 Sprat, *History*, p. 316.

46 For studies of *Micrographia* see John Harwood, 'Rhetoric and Graphics in *Micrographia*', in Michael Hunter and Simon Schaffer (eds), *Robert Hooke: New Studies* (Woodbridge, 1989), pp. 119–47; Michael A. Dennis, 'Graphic Understanding: Instruments and Interpretation in Robert Hooke's *Micrographia*', *Science in Context*, 3 (1989), 309–64; Janice Neri, *The Insect and the Image: Visualising Nature in Early Modern Europe, 1500–1700* (Minneapolis, 2011),

ch. 4; and Meghan C. Doherty, 'Discovering the "True Form": Hooke's *Micrographia* and the Visual Vocabulary of Engraved Portraits', *Notes & Records*, 66 (2012), 211–34.

47 Griffiths, 'John Evelyn and the Print', p. 110, and pp. 109–12 passim.

48 British Museum E,2.188. For Evelyn's account of Dürer, see Bell (ed.), *Evelyn's Sculptura*, pp. 37ff. (including pp. 38–9 on 'The Life of the Virgin'), 63.

49 For example, that by Antonello da Messina (c. 1475) in the National Gallery, London (this also has an architectural surround and chequered floor); this and other examples are reproduced in Dora Thornton, *The Scholar in His Study: Ownership and Experience in Renaissance Italy* (New Haven and London, 1997).

50 See Thornton, *Scholar in His Study*, esp. ch. 6.

51 Griffiths, 'John Evelyn and the Print', p. 111; Bell (ed.), *Evelyn's Sculptura*, p. 73; the Stradanus plates are conveniently reproduced in Susan Dackerman (ed.), *Prints and the Pursuit of Knowledge in Early Modern Europe* (Cambridge, MA, New Haven and London, 2011), pp. 38ff., or http://www.virtuelles-kupferstichkabinett.de/. For a discussion of the plate reproduced here (the precise date of which is unclear), see Ad Stijnman, 'Stradanus's Print Shop', *Print Quarterly*, 27 (2010), 11–29.

52 Reproduced in Domenico Bertoloni Meli, *Thinking with Objects: The Transformation of Mechanics in the Seventeenth Century* (Baltimore, 2006), p. 151. We are grateful to Domenico Bertoloni Meli for drawing this to our attention.

53 Evelyn, *Diary*, vol. 2, p. 230. On Kircher's cabinet see Paul Findlen, *Possessing Nature: Museums, Collecting and Scientific Culture in Early Modern Italy* (Berkeley and Los Angeles, 1994), and for the views on it of John Ray and Philip Skippon, Michael Hunter, 'John Ray in Italy: Lost Manuscripts Rediscovered', *Notes & Records*, 68 (2014), 93–109, on pp. 102–4.

54 Evelyn, *Diary*, vol. 3, pp. 110–11 (after 'Perspectives', Evelyn wrote 'places to introduce the *Species*', an allusion to Wilkins's work on his universal language). See also pp. 108 (items at St John's College) and 293 (Dudley Palmer's collection). For a commentary on Evelyn's account of Wadham, see Lisa Jardine, 'Dr Wilkins's Boy Wonders', *Notes & Records*, 58 (2004), 107–29.

55 For the Cracherode copy of the book, see British Museum 164.e.1, catalogued on Merlin (the British Museum database) as 1972, U.373.1 et seq. For secondary literature see esp. Dominique Jacquot, 'Le Grand Oeuvre de Chaperon. La "Bible de Raphael"', in Sylvain Laveissière, Dominique Jacquot and Guillaume Kazerouni (eds), *Nicolas Chaperon, 1612–1654/5* (Nimes, 1999), pp. 159–73, and *Raphael et l'Art Français* (Paris, 1983), pp. 195–6. For Evelyn's copy of the book, see Griffiths, 'John Evelyn and the Print', p. 110; Evelyn refers to it in Bell (ed.), *Evelyn's Sculptura*, p. 89.

56 See Jacquot, 'Le Grand Oeuvre', p. 165; *Raphael et l'Art Français*, p. 195; and *Raphael dans les Collections Françaises* (Paris, 1983), pp. 101ff.

57 The inscription on the column is as follows: 'Ille hic est Raphael, timuit quo sospite vinci rerum magna parens et moriente mori – non pulvis non umbra sumus me vivere Chapron hic dedit: Urbinas ille ego sum Raphael.' (the opening words come from Raphael's epitaph at the Pantheon). The inscription on the base is: 'N. Chapron Inventor' and 'Ad eximium pictorem Nicolaum Chapron, Picturam Raphael mutis revocavit ab umbris, tu Chaprone iterum reddis utrique diem'.

58 See Matthias Winner, 'Poussins Selbstbildnis von 1649', in Centre Nationale de la Recherche Scientifique (ed.), *'Il se rendit en Italie'. Etudes Offertes à André Chastel* (Rome, 1987), pp. 371–401, and Jacquot, 'Le Grand Oeuvre', p. 167.

59 Jacquot, 'Le Grand Oeuvre', p. 168. For the paintings in question see Richard Verdi, *Nicolas Poussin 1594–1665* (London, 1995), pp. 171–2, 176–7.

60 The book is Jacques de Bie, *Nomismata Imperatorum Romanorum Aurea* (1617). See J. Richard Judson and Carl Van de Velde, *Book Illustrations and Title-Pages (Corpus Rubenianum Ludwig Burchard, XXI)* (2 vols., London and Philadelphia, 1978), no. 39, vol. 1, pp. 188–93, vol. 2, fig. 130. It is interesting that this is a design for which the artist's drawing survives (fig. 131); this is now in the British Museum, 1900,0824.137, and it is very similar to the engraved version by Michel Lasne.

61 Jacquot, 'Le Grand Oeuvre', p. 167, and *Raphael et l'Art Français*, p. 195. The full dedication to Renard appears on the title-page.

62 The widespread circulation of such prints is perhaps suggested by Evelyn's comment in his translation of Roland Fréart, Sieur de Chambray's *An Idea of the Perfection of Painting* (London, 1668), concerning engravings like those by Marcantonio Raimondi that he might have used to illustrate that book: 'that such as are *Curious*, must needs already be furnish'd with them' (sigs b7v-8).

63 See Evelyn, *Diary*, vol. 2, pp. 113, 117, 120, 288, 295, 313, 357, 487, 495.

64 John Martin Robinson (ed.), *Remembrances of Things Worth Seeing in Italy Given to John Evelyn 25 April 1646 by Thomas Howard, 14th Earl of Arundel* (Roxburgh Club, 1987). On the portrait of Ferry Carondelet and attendants, now thought to be by Sebastiano del Piombo, see David Howarth, *Lord Arundel and His Circle* (New Haven and London, 1985), pp. 66–8, and Francis Haskell, *The Kings Pictures: The Formation and Dispersal of the Collections of Charles I and His Courtiers*, ed. Karen Serres (New Haven and London, 2013), p. 16. See also the Raphael *St Catharine of Alexandria* formerly owned by Arundel in Hollar's self-portrait, Fig. 3.8.

65 Bell (ed.), *Evelyn's Sculptura*, pp. 41–3; Evelyn, *Idea of the Perfection of Painting*, sig. A7 and passim. On Evelyn's translations of Fréart see also Hanson, *English Virtuoso*, pp. 83ff., and Thomas, 'Evelyn's Project of Translation'.

66 Evelyn, *Diary*, vol. 3, p. 287.

67 See Jane Roberts, 'The Limnings, Drawings and Prints in Charles I's Collection', in Arthur MacGregor (ed.), *The Late King's Goods* (London and Oxford, 1989), pp. 115–29, on pp. 117–18; Graham Reynolds, *English Portrait Miniatures* (revised edition, Cambridge, 1988), pp. 31ff.; and John Murdoch, Jim Murrell, Patrick J. Noon and Roy Strong, *The English Miniature* (New Haven and London, 1981), pp. 85ff. For a comment on Mary Evelyn's copy in this context see Edward Norgate, *Miniatura or the Art of Limning*, ed. Jeffrey M. Muller and Jim Murrell (New Haven and London, 1997), pp. 19–20.

68 Evelyn's purchase and the price he paid were specifically recorded by George Vertue: *Walpole Society*, 24 (1936) (*Vertue Note Books*, vol. 4), p. 196. For references by Evelyn to miniatures by Oliver and to Jerome Laniere, whose collection contained the one that Evelyn purchased, see Evelyn, *Diary*, vol. 2, p. 540, and vol. 3, pp. 74–5, 260.

69 For a recent account of the painting and particularly its frame see Christopher Rowell, 'The Kingston Lacy "Raphael" and Its Frame (1853–6) by Pietro Giusti of Siena', in *National Trust Historic Houses & Collections Annual* (2014), pp. 40–7.

70 For evaluations of Charles's cultural role see Kevin Sharpe, *Rebranding Rule: The Restoration and Revolution Monarchy, 1660–1714* (New Haven and London, 2013), ch. 2, and the briefer remarks of Antonia Fraser, *King Charles II* (London, 1979), pp. 192–3, 329–32, and Ronald Hutton, *Charles II, King of England, Scotland and Ireland* (Oxford, 1989), p. 450.

71 John Evelyn (trans), Roland Fréart, Sieur de Chambray, *Parallel of the Antient Architecture with the Modern* (London, 1664), sigs a3v-4.

72 Evelyn, *Idea of the Perfection of Painting*, sig b4v.

73 For a telling recent account of this process, see Matthew C. Hunter, *Wicked Intelligence: Visual Art and the Science of Experiment in Restoration London* (Chicago, 2013), ch. 6.

74 W. E. Knowles Middleton (ed. and trans.), *Lorenzo Magalotti at the Court of Charles II: His* Relazione d'Inghilterra *of 1668* (Waterloo, Ontario, 1980), p. 136. He also noted how Mary Evelyn, 'who was brought up at Paris, paints miniatures with great delicacy'; perhaps he had been shown the 'Raphael' copy.

75 See esp. Carol Gibson-Wood, 'Jonathan Richardson, Lord Somers's Collection of Drawings, and Early Art-Historical Writing in England', *Journal of the Warburg and Courtauld Institutes*, 52 (1989), 167–87, on pp. 182ff. For a more jaundiced view of the harmful effect that this attempted marriage of art and science had on artistic appreciation, see Bell's comments in Bell (ed.), *Evelyn's Sculptura*, pp. xvi ff.

76 On French developments see Antoine Schnapper, 'The Debut of the Royal Academy of Painting and Sculpture', in June Hargrove (ed.), *The French Academy: Classicism and Its Antagonists* (Newark, 1990), pp. 27–36; Donald Posner, 'Concerning the "Mechanical" Parts of Painting and the Artistic Culture of Seventeenth-Century France', *Art Bulletin*, 75 (1993), 583–98; Paul Duro, *The Academy and the Limits of Painting in Seventeenth-Century France* (Cambridge, 1997); see also Roger T. Hahn, *The Anatomy of a Scientific Institution: The Paris Academy of Sciences, 1666–1803* (Berkeley and Los Angeles, 1971), chs. 1–2. However, although the Académie de Peinture et de Sculpture bore some similarity to the Royal Society in its weak and underfunded position in its early years, in other respects (e.g., its teaching function and its championship of the intellectual over the technical aspects of painting) there are few resonances with the English state of affairs. For a further appraisal see Hanson, *English Virtuoso*, pp. 83ff.

77 *Phil. Trans.*, 3 (1668), 784–5, on p. 785. On this review see Hanson, *English Virtuoso*, pp. 21–2. For a review of Felibien's *Entretiens sur les Vies et sur les Ouvrages des plus Excellentes Peintures, Anciens et Modernes*, in which Raphael figures prominently, see *Phil. Trans.* 1 (1665–6), 383–4.

78 See Vera Keller, 'Scarlet Letters: The Mayerne Papers within the Royal Society Archives', paper delivered at the conference, 'Archival Afterlives', at the Royal Society, London, 2 June 2015. On Mayerne's activities see Hugh Trevor-Roper, 'Mayerne and His Manuscript', in Howarth, *Art and Patronage in the Caroline Courts*, pp. 264–93, and Trevor-Roper, *Europe's Physician: The Various Life of Sir Theodore de Mayerne* (New Haven and London, 2006), ch. 21. For the committee on painting see Birch, *Royal Society*, vol. 2, pp. 84, 107, 111, 202 and 227ff., including the letter from the artist, Alexander Marshall, printed on p. 231.

79 Birch, *Royal Society*, vol. 2, p. 229.

80 Sprat, *History*, p. 253. For a commentary see Hunter, *Wicked Intelligence*, p. 167.

81 Evelyn, *Idea of the Perfection of Painting*, sigs b1–2.

82 We are indebted to Jenny Millar for her advice on these matters.

4 Details

The portraits, books and institutional accoutrements of the Royal Society

General

At this point we need to return to the question of responsibility for the components of the print. As we have seen, there can be no doubt that Evelyn was responsible for its overall composition, though Beale's prompting may have played a subordinate role. On the other hand, when it comes to details, two points need to be made. The first is that, although the evidence surveyed in chapter 2 shows that when Evelyn prepared the design he intended it for Beale's project rather than Sprat's, the letter from Beale of 29 April 1667 discussed at the end of that chapter – together with the fact that the plate is actually dated 1667 – means that it is conceivable that it was only etched after the idea had been suggested of transferring it to Sprat's *History*. Hence, just as the 'particulars' of the society's scientific work that appeared in Sprat's book evidently owed something to the input of the society's organisers, this could also be true of the specimens of scientific apparatus and other accoutrements shown in the frontispiece: it is therefore not surprising that, as we will see in chapter 5, certain of the instruments that are depicted relate to projects that are divulged in Sprat's text. On the other hand, this is not exclusive of Evelyn having a particularly prominent role in deciding just what should be shown, so what we know of his views on such matters also needs to be canvassed.

It also seems likely that in the exact delineation of specific facets of the print – for instance the faces of the figures at the centre – the responsibility was neither that of Evelyn nor his Royal Society colleagues but that of the man responsible for etching the plate, Wenceslaus Hollar. It thus seems possible that Hollar was left to his own devices when it came to the depiction of certain objects that were included, most notably the society's mace (as we will see later in this chapter), and here he may have drawn on a repertoire of exemplars that were familiar to him. In addition, we need to make allowance for the fact that, even when he was trying to do justice to pieces of scientific equipment that would have been new to him, he may sometimes have omitted crucial details although his representation generally seems quite accurate.[1]

But even these possibilities are not mutually exclusive, since, on the basis of what is known about print-making practice at the time, we can visualise a degree of collaboration in the evolution of the design. Following Hollar's initial input in translating Evelyn's drawing into etched form, it is quite possible that he produced a proof which was shown to Evelyn and his Royal Society colleagues, on the basis of which modifications might have been made. Though proofs of Hollar prints are not common, examples do exist – for instance of his etching after a design by Francis Barlow for the frontispiece to John Ogilby's *Britannia* of 1675, now in the Royal Collection at Windsor, which lacks various details that appear in the final version, while certain components are sketched in ink.[2] This implies that changes might have been made between the print's initial design and its actual execution on the basis of consultation with his clients, though in the case of the Sprat frontispiece we can only speculate what these might have been.

The central figures

Turning now to the print's details, let us begin with the three portraits at its heart. These almost certainly reflect Hollar's well-known artistic skills, particularly in the delineation of their faces, since the evidence of Evelyn's artistic activity discussed in the previous chapter suggests that he would have been unable to emulate such sensitive portraiture. First, we have the figure of William, 2nd Viscount Brouncker, 'Societatis Præses', as the inscription beneath him states, and Brouncker had indeed been President of the society since he was so named in its first charter in July 1662.[3] In the case of Brouncker's figure, all that Hollar had to do was to add his facial features to a figure that is otherwise almost indistinguishable from that of Chaperon in the etching on which the print is based – even down to his shoes (see Fig. 4.1). Brouncker's face was probably taken from the life, and this is in fact the only known contemporary portrait print of Brouncker (though more than one oil portrait of him survives): this means, incidentally, that copies of the frontispiece have long been in demand by collectors of engraved portraits of the famous, one of the complications in the later history of the print that will be discussed in chapter 6.[4]

As for the portrait of Charles II in the form of a bust standing on the column at the centre of the print, again the facial features might have been executed by Hollar from the life. However, in this instance more changes were made from his exemplar. Apart from being reversed so that it faces right rather than left, the bust differs from that in Chaperon's print in various respects. The inner vest that Raphael wore under his toga (echoing the portraits referred to in the last chapter) is abandoned, and instead an elegant scarf is draped around Charles's shoulders.[5] More significant, whereas Raphael had long hair and a beard, Charles is shown clean-shaven apart from having a moustache, and he is also shown with cropped, curly hair, in a manner reminiscent of a Roman emperor. The latter feature is actually

Figure 4.1a and 4.1b Chaperon's self-portrait (Fig. 3.13) and the image of Lord Brouncker in the frontispiece: the robe, garter, stockings and shoes are identical.

one of the oddities of the print, since such a depiction of the king is almost unique at this stage in his career, when he was normally shown in prints and paintings wearing a flowing wig. Only in the 1680s were various equestrian and other statues made of him in which Charles was shown with cropped hair in the mode of a Roman emperor, and the evidence suggests that these were not well understood at the time.[6] On the other hand, an analogy is to be found in the way in which Charles was sometimes depicted on medals at an earlier date – something with which Evelyn would undoubtedly have been familiar, since his book, *Numismata, A Discourse of Medals Antient and Modern* (1697), illustrates how devoted he was to collecting such artefacts, long valued by the cognoscenti of Renaissance Europe (and thus providing him, incidentally, with a further apprenticeship in symbolic messages of the kind with which the Sprat frontispiece is concerned). *Numismata* includes various medals of Charles from the 1660s in which the monarch was shown (in Evelyn's words) 'in short Hair *á la Romain Antique*', including the Naval Reward Medal struck following the Battle of Lowestoft in 1665 (Fig. 4.2b).[7]

In fact, though the idea of using a bust echoed Chaperon's design, it is interesting to consider the significance of the depiction of the king in this manner. Just as Chaperon's statuesque depiction of Raphael denoted his pre-eminence in the realm of art, is Charles's rule being presented as part

Figure 4.2a and 4.2b The bust of Charles II in the frontispiece, compared with the wreathed image of him in 'short Hair and *Roman*-like' on the Naval Reward Medal of 1665.

of the natural order of things? In addition, is there a symbolic import in Charles's delineation in this manner? It is well-known that, although a visit to the Royal Society by Charles was planned in 1664, this failed to materialise and he never actually visited the society.[8] Hence his presentation in the form of a bust could be seen to denote his remote presence as patron, his disembodied, almost allegorical, representation elevating him to the sphere of the ideal. Is there a hint that his role in founding the society has elevated him to the realm of immortality, that his depiction in this manner almost represents his apotheosis? It might be added that, even more than the medals just referred to, Charles's bust is reminiscent of classical sculptures like the Belvedere Hermes at the Vatican. It is thus conceivable that this mode of presenting the king could be significant in itself in terms of the ambition of reviving antique models in art and architecture that was central to Evelyn's cultural programme.

Turning to the third figure, that of Francis Bacon, we must first briefly comment on the inscription beneath it, 'Artium Instaurator', 'Renewer of the Arts', a description of Bacon that is often cited.[9] This obviously echoes the title of Bacon's blueprint for science, *The Great Instauration*, and it also echoes various early eulogistic descriptions of him, for instance by the Bohemian reformer, Jan Amos Comenius, or the poet Alexander Gill, both of whom used the concept of 'the instauration of the arts' in connection with Bacon.[10] Similar sentiments occur more than once in the prefatory material to the 1640 Oxford edition of Bacon's *Advancement of Learning* – one of the books from which Evelyn made most extracts in his commonplace books – where Bacon's goal is slightly more fully described as being 'An Instauration of Sciences and Arts'.[11] The briefer formulation thus constitutes a perfectly appropriate label for Bacon, its limitation to two words possibly being partly dictated by the need for it to match the description of Lord Brouncker on the opposite side of the print.[12]

As for Bacon's depiction, this, particularly its facial features, seems to be based on the half-length portrait engraving of him made by Simon de Passe in 1618 when he was Lord Chancellor, including both his elaborate robes and the purse of office that he here ostentatiously holds in his right hand (this was omitted in later versions of the de Passe print, following his impeachment) (Fig. 4.3). In the portrait by de Passe, Bacon wears a hat, as was the case in many of the images influenced by it, including a later one by Hollar himself; here, however, this is omitted since it would obviously have looked strange in juxtaposition with the bare-headed portraits of Brouncker and Charles II.[13]

Bacon's body, including its rather indeterminate gesture, is not paralleled in the earlier portrait and is slightly problematic. The depiction of Bacon in his official robes was perhaps intended to denote his role as legislator of science, though obviously it also echoes the de Passe portrait. However, the

Figure 4.3a and 4.3b The depiction of Bacon compared with Simon de Passe's engraved portrait of 1618.

details of Bacon's costume that are visible in the two images are not identical, the sleeves, in particular, being noticeably different. As for the lower part of Bacon's body, this might possibly have been formed by reversing the depiction of Brouncker's legs on the other side of the print (though, in this case, they are largely concealed by Bacon's robes, apart from the not dissimilar shoe that protrudes). Bacon's pose is in fact rather ungainly, since his head faces towards the centre of the print, whereas his body is turned as far in the opposite direction as the need for an elegant posture will allow. It is even conceivable that the depiction of his body could have been lifted from some other image of him or a similarly clad figure which has not yet been identified. If this were the case, the hand gesture, which is reminiscent of the elegant gestures used in portraits by Anthony Van Dyck or William Dobson with which Evelyn and Hollar would have been familiar, could have come from the same source, perhaps being directed towards something included in the original that is missing here.[14] A further possibility is that Bacon's gesture is deliberately intended to direct our attention out of the picture, perhaps towards the title-page of the book which it was intended to face – whether Beale's or Sprat's.

On the other hand if, as is equally likely, the gesture is intended to have a significance within the current composition, there are two possibilities. One is that Bacon gestures towards the object to which he appears immediately to point, namely the gun, which could be significant in itself.[15] We will see in chapter 5 how, although (like the other instruments shown) the gun has a context in the Royal Society's activities in the period preceding the preparation of the frontispiece, it differs from the rest in not being a bespoke object but a standard gun of a kind that was commercially available. This being the case, it is conceivable that it could have an import rather different from the other items, perhaps acting as a reminder of Bacon's well-known emphasis in his *New Organon* on gunpowder (along with printing and the mariner's compass) as one of the inventions which had 'changed the face and condition of things' since antiquity (this is not, of course, exclusive of the gun also having a significance in terms of the Royal Society's programme).[16] It is equally possible, however, that Bacon's gesture is not to a specific object at all, but is intended to lead the eye up the right-hand side of the print more generally, thus creating a kind of circular movement through the design and linking its disparate components together. Whatever the case, it certainly enhances the print's air of pageantry.

The society's institutional accoutrements

We now turn to the details of the print which remind us of the great significance that was attached to the society's institutional role in its early years, and particularly to the accoutrements of its royal patronage. First, we have the society's coat of arms, as granted in its second charter of 1663, namely a shield with the three lions of England in the dexter corner but otherwise

Figure 4.4 The society's coat of arms with its accompanying theatrical swag of curtain.

blank, supported by two white hounds gorged with crowns and with a crest comprising a helmet with a crown on it surmounted by an eagle and with elaborate swags on either side. Under the coat of arms, on a ribbon, is the society's motto: 'Nullius in verba' (Fig. 4.4).[17] This is a standard depiction of the society's arms, very similar, for instance, to that which appears on the verso of the half-title of Sprat's *History* (Fig. 2.1), except that in this one the shield held in the eagle's right talon on which the royal lions are repeated is lacking. Here, it appears in the centre at the top of the print, flanked by curtains gathered with ribbons which are evidently intended to give dramatic effect.

Then, more than halfway down the left-hand side of the print, on a table draped with a cloth in front of the bookcase, there is a group of objects that again allude to the society's 'established' status (Fig. 4.5).[18] At the centre is

Figure 4.5 The symbols of the society's institutional status: charter, statutes, journal book and mace.

a document folded so that it is approximately square in shape, on which is written 'Diploma'; protruding from this is a seal on which what seems to be a coat of arms is just visible. This represents the elaborate charter that was drawn up for the society in 1662 and of which a revised version was issued a year later, to both of which the royal seal was attached and one of which this evidently depicts. Aligned with this but just beyond it is a bound volume inscribed: 'Stat:/ Reg:/ Soc:', i.e., 'Statuta Regalis Societatis', the society's statutes regulating its proceedings, approved in 1663, while on its side leaning up against the bookcase, with its bottom edge filling the space between the lowest shelf of the bookcase and the charter, is a comparable bound volume inscribed 'Iour/nal', in other words the society's journal book. The spine and fore-edge of the volume are visible and the implication is that this is the rear cover. Again, its prominent position bears witness to the significance attached by the society to the records that it kept of its proceedings.[19]

Lastly, lying across the charter, almost but not quite parallel with the word 'Diploma', is what is presumably intended to be the royal mace presented by Charles II in 1663, a further key symbol of the society's institutional standing, in that it was comparable to the mace borne before the speaker of the House of Commons. This impressive silver gilt object survives today and it continues to be valued for its symbolic role, though surprisingly little has been written about it since a seminal paper by C.R. Weld in 1846 in which he demonstrated that (contrary to popular belief) this was not the mace ejected from the House of Commons in 1653 but was specially commissioned for the society a decade later.[20] In this context, what is strange about the depiction of the mace that appears in the frontispiece is that it is not at all accurate. As Fig. 4.6 shows, the actual mace has a cylindrical shaft with a bulbous base and two globular protrusions, above the upper of which a pair of brackets support a substantial urn-shaped head embossed with various figures and other motifs and surmounted by a crown, ball and cross. By comparison, the object that appears in the frontispiece looks more like a wand of office, its stem expanding in profile so that it is slightly bulbous just above what looks like a ring of ivory or some similar material halfway down it. Its base is not visible, but its head is rather small compared with that of the actual mace, to which it bears very little resemblance: it seems to comprise an openwork crown, similarly headed with a cross but entirely lacking the embossed urn with its elaborate figures.[21]

This is one of the most puzzling features of the frontispiece. It seems unlikely that Evelyn would have mis-depicted the mace in this way, since in 1663 he had presented the society with a special cushion on which to place it, the gift of his father-in-law, Sir Richard Browne, and he must frequently have seen the mace itself at meetings of the society, where it was routinely displayed.[22] Surely we must here see the initiative of Hollar, who was perhaps told to include the mace but was not granted an audience with the actual object, and who therefore devised something that seemed to him more or less appropriate. It is perhaps worth noting that the implication is that any other

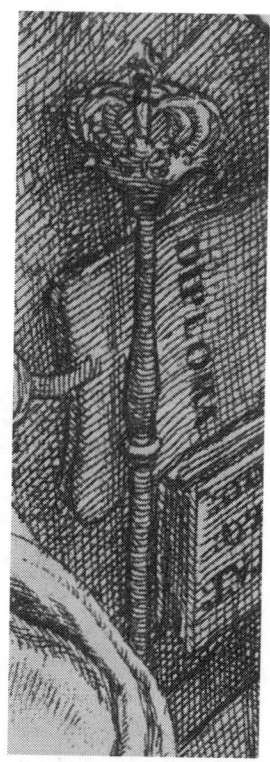

Figure 4.6a and 4.6b The mace as shown by Hollar in juxtaposition with the Royal Society's actual mace, presented by Charles II in 1663 and still extant today.

object depicted in the frontispiece could also be misleadingly depicted, and this needs to be borne in mind in what follows, though in general the level of detail suggests that Hollar was shown actual exemplars: perhaps the sheer value of the mace meant that it was the exception to this rule.

The books

Turning now to the books on the shelves above this group of objects (Fig. 4.7), these are as follows. There are four shelves of books, most of them standing upright but with some at a picturesque slant and some placed on their sides. In total, over fifty volumes seem to be shown, though it is not

Figure 4.7 The bookcase on the left of the print.

always clear where one book ends and the next begins, partly because they are shelved with their fore-edge rather than their spine protruding; in format, they evidently vary from folio to octavo or even duodecimo. A few of the books have lettering on their fore-edge, from which the intended identity of the volume can be deduced. The content of the shelves is as follows:

Top shelf. Most of the books stand upright, but in the foreground a couple are on their side, with a further book propped above them at an angle, leaning towards the upright books.

- One of the upright books, the second from the viewer, has on its fore-edge the word 'Harvey', alluding to William Harvey (1578–1657), the originator of the doctrine of the circulation of the blood who was often perceived as a key precursor of the Royal Society, and with whose anatomical investigations the society was naturally keen to associate itself.
- The book propped at an angle bears the title 'SILVA', that is, John Evelyn's *Sylva, or a Discourse of Forest Trees* (1664), the first book to be printed with the Royal Society's official imprimatur.

Second shelf. This has larger books in the centre and smaller books at either end, with other small ones on their sides stuffed into the shelf above them, the group further from the viewer being at a slight slant. Two of the large books in the central section have inscriptions:

- About the fifth one in is inscribed 'Copernicus', thus alluding to the figure who had started the astronomical revolution to which the Royal Society was heir and hence one of the society's principal precursors.
- The second one in is inscribed 'Pinax', possibly followed by an 'M', that is, *Pinax Rerum Naturalium Britannicarum* (1666) by the physician and Fellow of the society, Christopher Merrett.

Third shelf. Some books are upright, others leaning at an angle. The following have inscriptions:

- Approximately the seventh book from the end has the title, 'Augmen Scien.', alluding to Francis Bacon's *De Augmentis Scientarum* (1623), the revised Latin version of his *Advancement of Learning* (1605) that he prepared as part of his *Instauratio Magna*.
- A volume about as far further along the shelf as the previous one is from the end is inscribed, 'Hist. of Colour' [?]. This is almost certainly an allusion to Robert Boyle's *Experiments and Considerations Touching Colours* (1664).
- The volume leaning at an angle closest to the viewer is endorsed with the word, 'Benetada' (or 'Poenatada' or 'Henelato', or some variant on this). The significance of this is unclear.

Bottom shelf. About half of the shelf has books shelved upright; the other half, closer to the viewer, has a pile of books on their side; some of these are quite long, perhaps denoting folio format. Of the upright books, three have titles:

- The second from the end has 'Novum Organ', that is, Bacon's *Novum Organum* (1620), his chief methodological work, which was also part of his *Instauratio Magna*.
- The fourth from the end has 'Gilbert', evidently an allusion to William Gilbert, who, on the basis of his pioneering study of magnetism, *De Magnete* (1600), has sometimes been seen as the first English empirical scientist.
- The seventh from the end has 'Origin of F[?]'. The third word is difficult to decipher; it is an attractive possibility that it reads 'Forms', thus comprising a reference to Boyle's *Origin of Forms and Qualities* (1666; 2nd edn., 1667). However, it should be recorded that it could be 'Ea' or 'Ba'. The apparent 'll' that follows is in fact probably part of the hatching on the books.

By way of commentary, these volumes fall into the following categories. First, there are books by authors whom the Royal Society would have wanted to claim as its progenitors: Copernicus on the second shelf, Gilbert on the fourth, and Harvey on the top one. It is perhaps worth noting that the choice of authors might be seen as somewhat chauvinist, with only Copernicus's name being present to signal the major contribution of non-English thinkers to the revolution in knowledge to which the society was heir. How conscious this was is unclear, but what is less surprising is the presence of two works by Francis Bacon, perhaps the most important of all the society's precursors, his *De Augmentis Scientarum* (1623) and *Novum Organum* (1620), both forming part of his great but unfinished manifesto for the reform of science, the *Great Instauration*.[23]

The other books that are identified all seem to be associated with the Royal Society, though these represent a slightly more mixed bag. Perhaps most predictable are the works by the society's *doyen*, Robert Boyle, his *Experiments and Considerations Touching Colours* (1664) on the third shelf, one of the key works in which he used experiments to vindicate a mechanistic view of matter, while on the fourth we almost certainly have his *Origin of Forms and Qualities* (1666; 2nd edn., 1667), the anti-Aristotelian treatise in which he set out his corpuscularian theory of nature. It is perhaps also unsurprising to find on the top shelf John Evelyn's own *Sylva, or a Discourse of Forest Trees* (1664), the significance of which has already been outlined. A slightly more surprising choice is Christopher Merrett's *Pinax Rerum Naturalium Britannicarum* (1666) on the second, an alphabetical listing of plants, animals, minerals and fossils to be found in the British Isles in which the author had been encouraged by the Royal Society.[24] This work has been the subject of somewhat mixed evaluations by Canon Charles E. Raven and others, but it was clearly seen as exemplifying the society's rationale, and its author was quite active in the society's proceedings at this point.[25] Indeed, he actually presented a copy of the book to the society in January 1667, and it could have been because of its topicality that it seemed appropriate to include it.[26]

On the other hand, there is at least one surprising absence in the form of the principal book, other than Evelyn's *Sylva*, to receive the society's imprimatur in the years preceding the preparation of the frontispiece, namely Robert Hooke's *Micrographia* (1665) (Fig. 4.8).[27] This book was briefly noted in chapter 3 in connection with the arresting images that Hooke prepared for it using a microscope, and it is indeed notable that the microscope itself – perhaps the most famous instrument associated with the Royal Society in its early years – is absent from the Sprat frontispiece. As we will see in chapter 5, this could be because, in contrast to the instruments made specifically for the society's use that dominate the selection shown, the microscope was readily available for commercial purchase and this might have militated

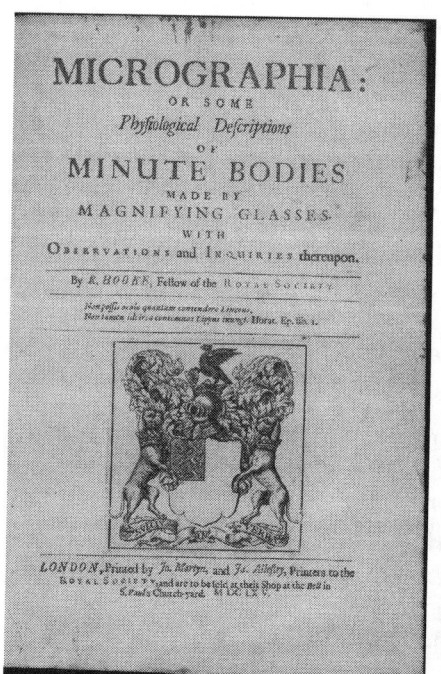
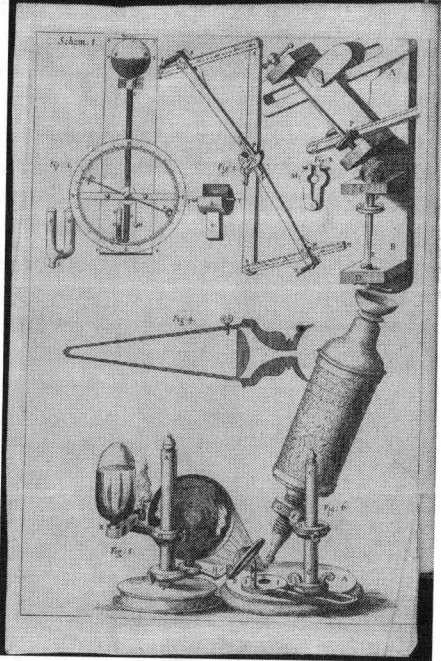

Figure 4.8a and 4.8b The title-page and opening plate of Robert Hooke's *Micrographia*, published with the Royal Society's imprimatur in 1665 but not among the titles shown on the shelves. Also conspicuously absent from the frontispiece to Sprat's *History* is the microscope shown in the plate, though the wheel barometer does appear: see p. 110.

against its inclusion. As far as the absence of Hooke's book is concerned, one possibility is that it was Evelyn who was responsible for choosing the titles that were included, and that this might reflect something of a blind spot on his part. In his *Diary*, he showed surprisingly little interest in the microscopic observations that Hooke demonstrated to the Royal Society in 1663 and 1664 and that were to be published in *Micrographia*, failing to share the excitement of contemporaries like Samuel Pepys about the hidden world that the microscope revealed, though he did show some interest in its potential for revealing the texture of natural things.[28] On the other hand, it is unclear whether this is an adequate reason for the omission of one of the most famous books to emanate from the society at this time – one which, moreover, was directly linked to its activities – and this remains a puzzling aspect of what would otherwise seem a fairly predictable batch of titles.

The background

The instruments in the background will be the subject of detailed comment in the next chapter. They hang from the walls of the architectural surround behind the figures at the centre of the print, or are laid out within it, and they are shown in perspective, so that their true size is generally apparent, though it is possible that in a few cases liberties were taken with the scale. Here it is worth repeating how architectural structures like this were a common feature of Renaissance religious art, and it seems unlikely that any particular significance should be attached to this one. Within Chaperon's etchings of the Raphael loggia, for instance, somewhat similar structures appear, and this is equally true of various book illustrations by Hollar, including his *Aesop*.[29] It is perhaps worth noting here that the relationship between the arches and pillars is not fully resolved in Hollar's treatment. Groins indicate that he intends the structure to be read as intersecting vaults but the arches springing towards the foreground from the rear pillars do not appear to end adjacent to the main transversal arch at the front.

   Turning now to the landscape background to the print, in general this seems a fairly typical vista of trees, hills and fields of the kind that appeared in many of Hollar's prints (there are also, of course, echoes of Evelyn's own earlier landscape etchings). Near to the telescope is what appears to be a ruined castle, and this too seems to be generic in form, though it bears a slight resemblance to a castle like Sheriff Hutton in Yorkshire, of which an etching was to be made by the Yorkshire virtuoso, William Lodge, in about 1680.[30] More striking is the edifice in the background on the right, which is so imposing that it seems likely that it is meant to have some significance. In its rural setting, this has the hallmarks of a Jacobean country house, with a central porch rising to three stories and flanking turrets, each with an elongated cap. Among surviving Jacobean mansions, it is most similar to Chilham Castle (Kent) (Fig. 4.9): but this may be coincidental, since the components that make up the composition are found in many of the show-houses built

Figure 4.9a and 4.9b The mansion in the background and its closest analogue, Chilham Castle in Kent.

in England in the first decades of the seventeenth century, including Blickling Hall (Norfolk), Charlton House (Greenwich) and Hatfield House (Herts.). Possibly, the aim was to allude to Bacon's own residence near St Albans, either Gorhambury, the Tudor house that he inherited from his father, Sir Nicholas Bacon, or the smaller structure, Verulam House, that Bacon himself erected (this had been demolished in 1663). On the other hand, the building shown here bears no close resemblance to contemporary drawings of the latter like that by John Aubrey.[31]

A further possibility is that it might have been intended to represent Chelsea College, the theological college established in the Jacobean period which had long been defunct and which Charles II granted to the Royal Society at this time: though the exact appearance of the college (which was demolished to make way for the Royal Hospital at Chelsea in 1681) is unclear, the building depicted in the frontispiece bears a vague resemblance to the idealised 'Modell' of it 'as it was intended to be built' that appears in John Darley's *The Glory of Chelsey Colledge Revived* (1662), which Evelyn or Hollar might conceivably have known.[32]

On the other hand, the edifice shown in the plate may not have been intended to represent a specific structure at all, but to evoke Solomon's House in Bacon's utopian sketch, *New Atlantis*.[33] For this, the general air of a Jacobean building would have been quite appropriate. Moreover it is worth stressing that, as depicted, the building is deliberately old-fashioned in style: a state-of-the-art structure of the 1660s would have looked quite different, like Thorpe Hall at Peterborough, for instance, among extant buildings, or, among lost ones, Sir Roger Pratt's highly influential Coleshill House (Fig. 4.10), which is clearly echoed in Evelyn's conception of an ideal college as outlined in the sketch accompanying his proposal to Boyle along these lines in 1659.[34] It can therefore definitely be stated that it cannot have been intended to represent the 'college' which the Royal Society sought funds to

Figure 4.10 Coleshill House, Berkshire. Sir Roger Pratt's masterpiece, destroyed by fire in 1952.

build in the Strand in 1667–8, the first reference to the aspirations concerning which appeared in the final pages of Sprat's *History*.[35] On balance, it seems likeliest that it represents some kind of intended allusion to Bacon.

Notes

1 For an example, see p. 100 in this book concerning his failure to show the winch, a crucial component of the mast for the telescope. See also pp. 90, 92.

2 RCIN 805074 (Pennington, *Hollar*, no. 2681, *New Hollstein*, no. 2332). On Hollar proofs, see Richard T. Godfrey, *Wenceslaus Hollar: A Bohemian Artist in England* (New Haven and London, 1994), p. 18, referring to Solomon and the Queen of Sheba (1642), British Museum 1870,0625.39 (Pennington, *Hollar*, no. 74, *New Hollstein*, no. 359) and Teresia, Lady Shirley (N.D.), British Museum 1861,0413.505 (Pennington, *Hollar*, no. 1503, *New Hollstein*, no. 2528). Another proof is Spring, from a series of allegories of the seasons (1643–4), Windsor, RCIN 802402 (Pennington, *Hollar*, no. 606, *New Hollstein*, no. 438): as will be seen, these are predominantly much earlier.

3 Birch, *Royal Society*, vol. 1, p. 90.

4 See James Granger, *A Biographical History of England* (2nd edition, 4 vols., London, 1775), vol. 4, p. 81, where his principal reference is to the Sprat frontispiece. However, in his account of portraits in oil, he confuses Brouncker with his brother, the 3rd Viscount: see David Piper (comp.), *Catalogue of Seventeenth-Century Portraits in the National Portrait Gallery 1625–1714* (Cambridge, 1963), pp. 35–6 (in connection with NPG 1567 and the portrait of Brouncker at the Royal Society).

5 For a bust with a slightly comparable scarf, see Hollar's etched title-page to Lovelace's *Lucasta* (1660), Pennington, *Hollar*, no. 2676, *New Hollstein*, no. 1743; bpi1700, no. 7287.

6 See Katharine M. B. Gibson, *"Best Belov'd of Kings": The Iconography of King Charles II*, Courtauld Institute PhD thesis, 1997, pp. 97ff. (the Sprat frontispiece is specifically dealt with on p. 100 and reproduced as plate 184). On the 1680s statues see also Nicola Smith, *The Royal Image and the English People* (Aldershot, 2001), pp. 125ff. On the iconography of Charles II as a whole see Kevin Sharpe, *Rebranding Rule: The Restoration and Revolution Monarchy, 1660–1714* (New Haven and London, 2013), ch. 2 (including pp. 140–2 on the Sprat frontispiece).

7 John Evelyn, *Numismata: A Discourse of Medals, Antient and Modern* (London, 1697), p. 128. Cf. p. 134 ('short Hair and *Roman*-like') and pp. 130, 136, 140. For the background see Francis Haskell, *History and Its Images: Art and the Interpretation of the Past* (New Haven and London, 1993), esp. ch. 1. For the Naval Reward Medal by Jan Roettier see Edward Hawkins, *Medallic Illustrations of the History of Great Britain and Ireland to the Death of George III*, ed. A. W. Franks and H. A. Grueber (2 vols., London, 1885), vol. 1, p. 503 (no. 139).

8 For the preparations, see Birch, *Royal Society*, vol. 1, pp. 288–91, 316, 340–1, 389, 391, 408.

9 For an example see n. 15 in this chapter. It is also striking how prominent the frontispiece is in the results of a Google search for this phrase.

10 See J. A. Comenius, *Pansophiæ Diatyposis* (Amsterdam, 1645), p. 202; W. D. Cooper, 'Unpublished Poems of Alexander Gill the Younger', *Gentleman's Magazine*, 35 (1851), 345–7, who refers to Gill's verses, 'Viro nobilissimo Francisco Baroni de Verulamio, Vicecom. S. Albani, novam magnamq. artium instaurationem gratulator', on p. 346 (these are in Frewen MS 690 at the East Sussex Record Office). We are indebted to Vera Keller for these references.

11 Francis Bacon, *Of the Advancement and Proficience of Learning* (Oxford, 1640), p. 2. Cf. also sig. C1 (a poem in 'Manes Verulamiani' by 'H. T. Coll. Trin. Socius' which contains the passage: 'At nostras etiam *Verulamius* artes INSTAURAT veteres, condit & ille novas'). For Bacon's *Advancement* in Evelyn's commonplace books see Michael Hunter, 'John Evelyn in the 1650s: A Virtuoso in Quest of a Role', in Hunter, *Science and the Shape of Orthodoxy* (Woodbridge, 1995), pp. 67–98, on p. 74. Evelyn's copy of this book is now British Library Eve.b.16.

12 Matters are slightly complicated by a further usage of which Evelyn may or may not have been aware, by Robert Burton in his famous *Anatomy of Melancholy* (1621) (not a book that appears in Evelyn's commonplace books). In 'Democritus to the Reader', the slightly longer phrase, 'Omnium artium & scientiarum instaurator' (echoing that in Bacon, *Advancement*, p. 2) is applied, not to Bacon, but to a different and more surprising figure, namely the mythical figure of Elias Artista, promoted by the sixteenth-century iatrochemist, Paracelsus, as the harbinger of the improved knowledge that would mark the end of time. See Robert Burton, *The Anatomy of Melancholy*, ed. T. C. Faulkner, N. K. Kiessling and R. L. Blair, with an introduction by J. M. Bamborough (6 vols., Oxford, 1989–2000), vol. 1, p. 108; vol. 4, pp. 163–4, and, on Elias Artista, Herman Breger, '*Elias Artista* – A Precursor of the Messiah in Natural Science', in Everett Mendelsohn and Helga Nowotny (eds), *Nineteen Eighty-Four: Science between Utopia and Distopia* (Dordrecht, 1984), pp. 49–72, and William R. Newman, *Gehennical Fire: The Lives of George Starkey, an American Alchemist in the Scientific Revolution* (Cambridge, MA, 1994), pp. 3–4, 11–12, 68. It is perhaps also worth noting that we have located yet another usage of 'Artium Instaurator' in *Petri Rami Vita* (1599) by the French doctor and savant, Nicolas de Nancel: see Peter Sharratt's edition of this work in *Humanistica Lovaniensia*, 24 (1975), 161–277,

on pp. 220–3. However, in view of Bacon's notorious disdain for Ramus (see, e.g., Benjamin Farrington, *The Philosophy of Francis Bacon* (Liverpool, 1964), pp. 63–4), this seems even less likely to be Evelyn's source.

13 For a helpful conspectus of early portraits of Bacon see http://www.npg.org. uk/collections/search/person/mp00201/francis-bacon-1st-viscount-st-alban?search= sas&sText=Bacon%2C+Francis&OConly=true. For Hollar's later portrait, exe-cuted for the 1676 edition of *The History of the Reign of Henry VII*, see Penning-ton, *Hollar*, no. 1355; *New Hollstein*, no. 2349. Hollar also did a rather different full-length etching of Bacon's effigy on his tomb at St Albans, in which Bacon is accoutred in a hat and fashionable clothing (Pennington, *Hollar*, no. 2243; *New Hollstein*, no. 2140). For a full-length portrait of Bacon in his Chancellor's robes, see the frontispiece to Richard Read's *Secrets of Art and Nature* (1660), though he is shown standing rather than sitting (and wearing a hat), and lacks his purse of office; this is by Richard Gaywood, almost certainly after Francis Barlow (attribution by Nathan Flis).

14 On this feature in van Dyck and Dobson see the Tate Britain Van Dyck exhibi-tion, ed. Christopher Brown and Hans Vlieghe, 2009, and the National Portrait Gallery Dobson exhibition, ed. Malcolm Rogers, 1983.

15 One author who had no doubt about this was J.R. Hale, 'Gunpowder and the Renaissance: An Essay in the History of Ideas', in his *Renaissance War Studies* (London, 1983), pp. 389–420, on p. 405, where he writes in relation to the Sprat frontispiece: 'Gunpowder manufacture was one of the industries that Francis Bacon recommended for study, and it is no accident that on the title page of Bishop Thomas Sprat's *History of the Royal Society*, Bacon, "Artium Instaura-tor," is portrayed pointing to a gun.'

16 Francis Bacon, *The New Organon*, ed. Lisa Jardine and Michael Silverthorne (Cambridge, 2000), p. 100. Yet another possibility, since, as we will see, the gun seems to be a sporting one, has been suggested to us by Nathan Flis. In his writ-ings, Bacon used the hunt as a metaphor for discovery and hence for the empiri-cal enterprise to which the society was devoted: this detail of the print might therefore be supposed to denote this Baconian concept of 'science as a *venatio*'. See esp. William Eamon, *Science and the Secrets of Nature: Books of Secrets in Medieval and Early Modern Culture* (Princeton, 1994), ch. 8, pp. 283ff.

17 See *The Record of the Royal Society* (4th edition, London, 1940), pp. 20–1, 238–9, 252, and plate 4. Cf. Hunter, *Establishing*, p. 17, and p. 28 in this book.

18 See Hunter, *Establishing*, esp. ch. 1; Mordechai Feingold, 'Of Records and Gran-deur: The Archive of the Royal Society', in Michael Hunter (ed.), *Archives of the Scientific Revolution: The Formation and Exchange of Ideas in Seventeenth-Century Europe* (Woodbridge, 1998), pp. 171–84.

19 See Feingold, 'Of Records and Grandeur'.

20 C.R. Weld, 'History of the Mace Given to the Royal Society by King Charles the Second', *Abstracts of the Papers Communicated to the Royal Society of London*, 5 (1843–50), 611–19, reprinted almost verbatim in Weld, *The History of the Royal Society* (2 vols., London, 1848), vol. 1, pp. 150–65. See also *The Record*, pp. 13–14 and plate 3.

21 It is perhaps worth noting that wands of office with slightly similar heads com-prising openwork crowns appear in the top corners of the de Passe portrait of Bacon that Hollar apparently used as a source: see p. 63.

22 Birch, *Royal Society*, vol. 1, p. 250. For his record of the arrival of the mace see Evelyn, *Diary*, vol. 3, p. 332, though, as de Beer points out, this entry is anoma-lously early in relation to the actual delivery of the mace in 1663.

23 It might have been expected, by way of analogy with the instruments dealt with in chapter 5, that it would be possible to identify the actual books whose titles

are inscribed on volumes in the bookcase as being in the society's possession. Certainly, the society had presentation copies of the works by Boyle and Merrett that appear: see William Perry (comp), *Bibliotheca Norfolciana* (London, 1681), esp. 'Libri ab Ipsis Autoribus et Aliis R. Societati Donati', pp. 169ff. (where the Boyle and Merrett items appear respectively on pp. 171 and 173; though *Sylva* appears on p. 170, this is the second edition of 1670, but there is also an 'Idem' which might represent that of 1664). Otherwise, however, the authors and titles do not tally well with the society's holdings: the society does not seem to have owned any work by Copernicus or Gilbert at this time, and the holdings by Bacon and Harvey were perfunctory (see esp. p. 14). It is perhaps worth adding that, since these seem to be 'notional' rather than real volumes, it is unsurprising that there is no clear correlation between their apparent size here and the actual format of the books involved.

24 Birch, *Royal Society*, vol. 1, pp. 138, 367.

25 For evaluations see C. E. Raven, *English Naturalists from Neckam to Ray* (Cambridge, 1947), pp. 305–38; A. J. Koinm, 'Christopher Merret's Use of Experiment', *Notes & Records*, 54 (2000), 23–32. See also C. E. Raven, *John Ray: Naturalist* (2nd edition, Cambridge, 1950), pp. 77–8, and R.W.T. Gunther (ed.), *Further Correspondence of John Ray* (London, 1928), p. 112. For Merrett's activity see e.g., Birch, *Royal Society*, vol. 2, pp. 165, 171.

26 Birch, *Royal Society*, vol. 2, p. 138. For Boyle's presentation of *Colours* in March 1664 see vol. 1, p. 404; for his presentation of *Forms and Qualities* in March 1666 see vol. 2, p. 76. For Evelyn's presentation of the second edition of *Sylva* in Dec. 1669 see vol. 2, p. 411; for the society's role in the inception of the book see ch. 3, n. 8 in this book.

27 For a list of books published with the society's imprimatur, see C. A. Rivington, 'Early Printers to the Royal Society, 1663–1708', *Notes & Records*, 39 (1984), 1–27, on pp. 22–7, with 'Addendum' in ibid., 40 (1986), 219–20. Apart from *Sylva* (1664) and *Micrographia* (1665), two editions (not three, as incorrectly stated by Rivington) of John Graunt's *Natural and Political Observations . . . upon the Bills of Mortality* were published with the society's imprimatur in 1665, as was Walter Charleton's *Inquisitiones II Anatomico-physicæ*, but the omission of these seems more understandable.

28 The only reference to Hooke's demonstrations is under 1 July 1663, when Evelyn records the order to print them (Evelyn, *Diary*, vol. 3, pp. 356–7, though this does not tally with the minutes, as de Beer notes). On the other hand, he ignores Hooke's microscopic presentations which had occurred at regular intervals since April that year (see John Harwood, 'Rhetoric and Graphics in *Micrographia*', in Michael Hunter and Simon Schaffer (eds), *Robert Hooke: New Studies* (Woodbridge, 1989), pp. 119–47, on pp. 124–5), although he was often present at the meeting in question. Later, Evelyn mentions microscopic presentations in *Diary*, vol. 3, p. 629, and vol. 4, pp. 126, 128 (though at this point, too, there were more such presentations than he notes: see Birch, *Royal Society*, vol. 3, pp. 352, 358, and Evelyn was certainly present on the second occasion: cf. *Diary*, vol. 4, p. 125). On the other hand, Evelyn *does* quote *Micrographia* on the structure of petrified wood in *Sylva* (London, 1664), pp. 96–7 (see Hooke, *Micrographia* (London, 1665), pp. 108–9), his notes on the endpaper to his copy of *Micrographia*, sold at Christie's in 1977, suggesting that his primary interest was in trees, plants and vegetables (see Christie's catalogue, *The Evelyn Library, Part II: D–L*, 30 November – 1 December 1977, lot 775). Evelyn refers to the microscope in his letter to Cowley cited in ch. 3, n. 11 in this book, and in *Elysium Britannicum*, ed. John E. Ingram (Philadelphia, 2001), pp. 76, 300–1, 309–10 (mainly in marginal addenda), while he speaks of his own use of a microscope in

connection with analysing types of soil in his *A Philosophical Discourse on Earth* (London, 1676), pp. 32ff. For Pepys's contrasting excitement, especially at the depictions of tiny animals for which *Micrographia* is famous, see his *Diary*, ed. Robert Latham and William Matthews (11 vols., London, 1970–83), esp. vol. 5, pp. 48, 235; vol. 6, pp. 17–18.

29 See in this book, pp. 40–1, 43–4 and ch. 3, n. 38. For the suggestion that 'the scene was perhaps suggested by the tiled piazzas on the north and south sides of the Green Court at Gresham College', see Keynes, *John Evelyn*, p. 284. However, this is not very plausible as the resemblance is not at all close.

30 See bpi1700, no. 1068, which tabulates the various impressions in the British Museum. Keynes, *John Evelyn*, p. 284, discerns 'a suggestion of Windsor Castle'. For a magnified view see Fig. 5.17.

31 For Aubrey's drawing of Verulam House see his *Brief Lives*, ed. Kate Bennett (2 vols., Oxford, 2015), vol. 1, p. 213; for his commentary both on this and on Gorhambury, see pp. 211ff.

32 John Darley, *Glory of Chelsey Colledge Revived* (London, 1662), frontispiece. This was redrawn for Francis Grose's *Military Antiquities* (2 vols., London, 1786–8), vol. 2, facing p. 181, and is reproduced from there in M. H. Cox and Philip Norman (eds), *Survey of London, vol. 11, The Parish of Chelsea, Part 4* (London, 1927), plate 2. See also Cox and Norman, pp. 1–4, and Thomas Faulkner, *Historical and Topographical Description of Chelsea and Its Environs* (2 vols., London, 1829), vol. 2, pp. 218–34. Both accounts cite a description of the building as it existed in 1652 which shows that it bore little relationship to Darley's print. The transfer of the college to the society was complex, being finally ratified in 1669.

33 For this suggestion, see Keynes, *John Evelyn*, p. 284, echoed in Pennington, *Hollar*, no. 459. It is perhaps worth pointing out that the building cannot be intended to depict Gresham College, depictions of which before it was demolished (such as George Vertue's of 1739, reproduced in Michael Hunter and Simon Schaffer (eds), *Robert Hooke: New Studies* (Woodbridge, 1989), p. 286) show that its profile was quite different.

34 Reproduced in Hunter, *Establishing*, p. 183. For Coleshill, see Nicholas Cooper, *The Jacobean Country House: From the Archives of Country Life* (London, 2006), pp. 182–9; this book also provides details of the other houses mentioned here.

35 See Sprat, *History*, p. 434; Michael Hunter, 'A "College" for the Royal Society: The Abortive Plan of 1667–8', in Hunter, *Establishing*, ch. 5, passim. See Hunter, p. 172n., concerning a drawing that was at one time thought to be Wren's design for this building (and is reproduced in *Oldenburg*, vol. 4, plate 4), though this is almost certainly mistaken. A final possibility that may be disposed of here is that the building was intended to depict Arundel House in the Strand, where the society was at this point holding its meetings: for a depiction by Hollar of Arundel House and its urban milieu see Hunter, p. 165 (and Pennington, *Hollar*, no. 1002; *New Holstein*, no. 1730).

5   The instruments

Jim Bennett

Introduction

The instruments assembled in the frontispiece repay individual scrutiny. On the one hand, the known circumstances of the production of the etching add to our knowledge of these objects, for it soon becomes evident that they were real instruments, not merely projections or proposals. On the other hand, the nature of the objects and the overall character of the selection add to what can be said about the project of making the frontispiece. Since there is evidence that the great majority of these instruments were real, it is tempting to extend that assumption to them all. No doubt there are inaccuracies in depiction at this small scale, but evident inconsistencies in how they are rendered do not mean that an object itself was wishful or imaginary. In the one case where an object in the frontispiece has survived to the present day, namely the society's mace, we know that an inaccurate depiction sits alongside its historical existence. Perhaps real objects were obligated by the motto hanging above them all: 'Nullius in Verba'.

These instruments were not chosen casually and, at the risk of beginning with the conclusions of this attempt to identify as many as possible, it may be helpful to have the main character of the selection in mind when looking at the individual examples. In addition to being real objects, they were generally involved with the activities of the society. For understanding the work of making the plate, it seemed important to assess how each instrument might have been encountered by the chief agents, in particular Evelyn and Hollar. This proved to be fairly straightforward: the instruments were in London, and most were even, at least at times, in Gresham College. Hooke emerges as the Fellow most implicated in the instrumental content, not surprisingly, perhaps, since he was Curator of Experiments. The period of their active use is also quite focused, for the most part clustering between the years 1664 and 1667.

Hooke's appointment as the society's Curator was a drawn-out affair, beginning with a proposal, unanimously accepted, from Sir Robert Moray in November 1662, but regularised as an official, salaried office only after a series of steps that began in July 1664.[1] During this process, Hooke moved into the college at the beginning of September, though he became Professor of Geometry only the following year.[2]

Figure 5.1 The central area of the frontispiece, selected to include all the instruments, keyed with the numbers used to identify them in this chapter.

Evelyn was directly involved with the society's repository at Gresham College in the core period for the design of the print. As we saw in chapter 3, on 21 March 1666 he was appointed to a committee 'to take care of the well ordering, preserving, and increasing the stock of the said repository'.[3] The committee included Hooke and was to meet the following Monday 'in Mr. Hooke's lodgings, continuing the same from time to time on that day, and in that place'. The fact that Hooke lodged in the college resulted in a seamless confusion between instruments relevant to his work and the contents of the repository. Evelyn records in his *Diary* that one reason for his going to London on 2 April 1666 was to 'consult about ordering the natural rarities belonging to the Repositorie of the R: Soc: referr'd to a Committè'.[4] Nor was this group inactive as interest in the repository grew. At a Council meeting on 4 June 1666, three members of the committee, Daniel Colwall, William Harrington and John Graunt, each of them with links to the City, were authorised to speak to the committee of the Mercers' Company responsible for Gresham College and ask 'that they would please to repair the floor and windows in the west gallery of the said college, where the society's repository is to be'.[5]

As for the character of the instruments, it is clear that they were selected to represent the innovative, active and (as it seemed to those involved) significant work of the society. There is a mixture of experimental philosophy and practical mathematics, which reflects the record of the minutes. At the same time, there is an absence of other subjects that interested the Fellows, such as husbandry or rarities. Most of these criteria for inclusion may seem self-evident, but they are not. A more rhetorical approach could have been adopted. The print could have been filled with the promise of the society's programme, rather than its reality. Instruments might have been selected for their extravagance or their elaborate appearance. There is extravagance in a 60-foot telescope and, in the seventeenth century, prestige in a novel form of clockwork, but these instruments too are present because they engaged the society's recent and current activity.

The instruments and other objects have been numbered, for reference, in Fig. 5.1, moving first down the left side of the image and then the right, and for the most part we shall respect that sequence in what follows. Occasionally, however, objects can helpfully be grouped together, even if they are dispersed in the image.

A longitude timekeeper

Instrument number 1 is one of the sea-clocks or longitude timekeepers devised by Alexander Bruce, Earl of Kincardine in the early 1660s.[6] Following an experimental short-pendulum longitude clock, designed by Bruce in 1661 and discussed with the Dutch mathematician Christiaan Huygens, collaboration between the two men in the ensuing year resulted in a design for a spring-driven clock, with a pendulum of seven inches, housed in a triangular case. Bruce paid for two such clocks to be made by Severyn Oosterwyck of The Hague, and brought them to England in December 1662. One was damaged on the voyage and replaced by a similar clock by the London maker John Hilderson. The Oosterwyck and Hilderson clocks were tried at sea by Captain Robert Holmes in April 1663, on a voyage to Lisbon, and they returned to London in September. Hilderson made some changes to his clock, and in November both went with Holmes to Gambia, returning in December 1664. In 1664 clocks of the same type, though with technical modifications, were made by other clockmakers in London, including John Fromanteel and Edward East, though we cannot assume they had triangular cases. In recent years two clocks from this series have re-emerged, having survived by being adapted to other timekeeping uses (Figs. 5.3, 5.4).[7]

Figure 5.2 Instrument number 1.

Figure 5.3 Marine timekeeper, signed by Severyn Oosterwyck, after the design of Alexander Bruce and Christiaan Huygens, made c.1662, in a later case and adapted to serve as a mantel clock. Private collection.

Figure 5.4 The movement of an unsigned marine timekeeper from the 1660s, after the design of Alexander Bruce and Christiaan Huygens, later converted to a weight-driven longcase clock. In the collection of the National Maritime Museum, Greenwich, object number ZBA6944.

The Royal Society took an active interest in these developments and Moray, Hooke and Bruce (at least) embarked on a short sea voyage in late February 1663 to test Bruce's clocks.[8] Hooke wrote a report of the trial, where two clocks were suspended beneath the deck of a ship 'by a Ball and Socket of Brass'.[9] This type of gimbal suspension, where a device was hung from a steel ball, supported by but free to rotate in a brass cylinder – a version of the 'Cardano' suspension (named after the sixteenth-century mathematician, Girolamo Cardano) – is evident in the frontispiece. This trial preceded those conducted by Holmes, but Hooke was writing after at least Holmes's first voyage.

The society maintained an interest in the trials undertaken by Holmes.[10] Oldenburg published a favourable account of the performance of the time-keepers on the voyage to West Africa in the first issue of *Philosophical Transactions*, for March 1665, pointing out that both Bruce and Huygens were Fellows of the society and that the trial had been arranged by 'some of our Eminent *Virtuosi*, and Grand Promoters of Navigation'.[11] It was also in March 1665 that a new type of pendulum clock for use at sea was included in the patent for various inventions that the Royal Society obtained, noted in connection with the activity of the Mechanical Committee in chapter 2.[12] Hooke was still active in the subject close to the preparation of the frontispiece and on 29 August 1666 'produced also a new piece of watch-work of his contrivance, serving to measure time exactly both by sea and land'.[13]

The most puzzling feature of the depiction in the print is that the time-keeper may appear to have only one hand, which of course was not the case. While this is not the only possible reading of the image,[14] a single hand would not compromise the identification but might question the accuracy of the print in every detail. We need not accept the suggestion that Evelyn's ignorance of horology was such that he thought a single hand was correct;[15] instead, the discrepancy may help attune us to the character of the print as historical evidence.

A beam balance

Instrument number 2 is evidently an equal-arm beam scale. At first sight this might seem to be a fairly ordinary set of scales, possibly connected with the chemical apparatus we shall come to later. In fact it is located at some distance from the chemistry group in the composition. On its own an 'ordinary' balance would be anomalous; we shall find as we work through the instruments, as we have noted for instrument number 1, that they represent innovation in recent or continuing activity at the society, so it would be odd to find a relatively everyday object so deliberately placed on top of the bookcase. In fact it is not ordinary at all. Balances from the period are not usually carried on pillar stands in this way, but would be suspended by hand when in use, while larger balances for grain, for example, would be suspended from

Figure 5.5 Instrument number 2.

a stand. The other notable feature is the length of the beam and its plain, unadorned design; it gives the impression of a carefully made beam intended for serious and heavy work. The use of heavy weights is implied also by the fact that chains are shown supporting the substantial pans, rather than cords. Another feature that places the instrument out of the ordinary is the use of three-line suspensions for the pans rather than four.[16]

In September 1664 Hooke, Brouncker and Moray were conducting experiments at the top of the tower of Old St Paul's Cathedral.[17] One involved seeing whether there was a difference in the weight of a body at the top and the bottom. Hooke himself had tried a similar experiment in 1662 at the top of Westminster Abbey.[18] There he had used 'a pair of exact scales and weights' to weigh a piece of iron and a length of 'packthread' at the top, at seventeen ounces and thirty grains troy. Letting it down by the packthread, he found it heavier by 'somewhat more than ten grains'.[19]

The experiment at St Paul's in 1664 was altogether more ambitious.[20] Here the experimenters had 'a very curious beam' and two weights, each of fifteen pounds; from the context in Hooke's report, he is almost certainly referring to weight in avoirdupois. The two substantial loads, one a single brass weight, the other a bag with smaller weights and some string, were first brought to equilibrium and the beam then 'hung . . . over the very middle of the steeple' and the bag of weights lowered to the floor. After 'long adjusting' they concluded that the lower weight was now lighter by 'a dracm' (a dram), which they attributed to the greater density of the air at the bottom. Hooke was keen to emphasise the sensitivity of his beam, the difference being

> very observable, that though the weight hung at that distance, and though, by some misfortune, the cock of the beam was missing, yet was the beam so tender, that a very small weight, as some very few grains, would sensibly turn it, and when brought to an equilibrium, the beam would vibrate, as if it had only a pair of short scales hanging to it.[21]

The 'cock' or pointer is also missing from the beam in the frontispiece. The trefoil visible in the print is very unusual and might be some representation of the suspension arrangement. From Hooke's account, the weights were first balanced with the beam, before the latter was hung over the crossing of the cathedral, so a pillar stand appropriate to a substantial beam may also have had a part to play. Finally, we might imagine that the curious placing of the instrument on top of the bookcase alludes to its elevated role in the experiment.

Measuring angles at a distance

In 1674 Hooke made an emphatic claim in his *Animadversions on the First Part of the Machina Coelestis of . . . Johannes Hevelius* that his record of designing instruments for taking angular measurements at a distance, whether for astronomy, navigation or surveying, demonstrated innovation, versatility and fecundity:

> I have my self thought of, and in small modules try'd some scores of ways, for perfecting Instruments for taking of Angles, Distances, Altitudes, Levels, and the like, very convenient and manageable, all of which may be used at Land, and some at Sea, and could describe 2 or 3 hundred sorts . . . yet everyone differing one from another in some or other circumstantial and essential part.[22]

He had command of a number of technical elements, such as the telescopic sight, tangent screw adjustment, micrometer screw division, the bubble level, using mirrors to present images together, bringing images into coincidence by reflection so as to measure their angular distance, and so on, and he combined these in a range of instrument designs. Sprat himself, in the section on 'The Instruments they have invented' in his *History*, though without naming Hooke, grouped this range of instruments together, citing, inter alia, three different designs of portable quadrants for celestial or terrestrial measurement, 'manageable in any Window, or Turret, [they] are yet far more exact, than the best, that have been hitherto us'd'.[23] Sprat seems to have caught the tenor of Hooke's language and four of the instruments in the frontispiece (numbers 3, 5, 13 and 17) fall into this class. They can all be linked to the contemporary work of Robert Hooke.

Number 3 is a reflecting instrument for measuring angles between distant objects, especially for celestial navigation at sea. Sprat includes a rather precise reference to such an instrument:

> A new *Instrument* for taking Angles by reflection; by which means the Eye at the same time sees the two Objects, both as touching in the same point, though distant almost to a Semicircle: which is of great use for making exact *Observations* at Sea.[24]

This is the earliest published reference to such an instrument, whose principle of reflection became fundamental to navigational measurement in the

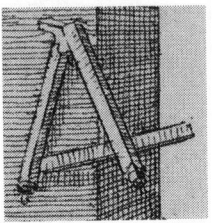

Figure 5.6 Instrument number 3.

octant and sextant, and the illustration at number 3 is the first published depiction.[25]

The two arms pivot on a shared hinge. The detailed mounting of the straight, graduated rule is not shown but it must be secured at a pivot on the left-hand arm, which is a handle, and pass through a slot or guide with an index on the right, which is a telescopic sight. This (together with the other instruments in the group) may be the earliest instance of a telescopic sight on a handheld instrument.

At a meeting of the society on 22 August 1666, Hooke 'mentioned a new astronomical instrument for making observations of distances by reflection, and was desired to give order for the construction of it, and to produce it before the society'.[26] The following week he mentioned that the instrument was 'for observing the positions and distances of fixed stars from the moon by reflexion', which would place it in the context of the lunar distance method for finding longitude; he 'was desired to have it made with speed'.[27] No meeting was held the subsequent week, on account of 'the late dreadful fire', but on 12 September Hooke 'presented his new perspective for taking angles by reflexion'.[28]

In use, one target would be sighted directly through the telescope, and the arm carrying the reflector (probably a mirror) was moved to bring the second target into apparent coincidence with the first. The angle between them would be registered by the straight rule. It would be rash to try to interpret too closely the apparent assembly of optical components around the hinge, but a mirror seems likely, possibly a reflecting prism, possibly a shade or filter, if solar observation was envisaged. Other priorities must quickly have preoccupied Hooke. At the following meeting he 'shewed his model for rebuilding the city to the society';[29] since the society heard that the plan had the support of the Lord Mayor and the aldermen, there must have seemed a real possibility of its being adopted. There was no further reference to Hooke's instrument at their meetings.

Hooke's *Posthumous Works* (1705), however, contains an account of an instrument working on this principle (see Fig. 5.7), which the book's editor Richard Waller says he found 'describ'd upon a loose Paper'; he placed it in the context of the lunar method for longitude.[30] Here the straight rule had linear graduations, to be converted into degrees by a table of chords.

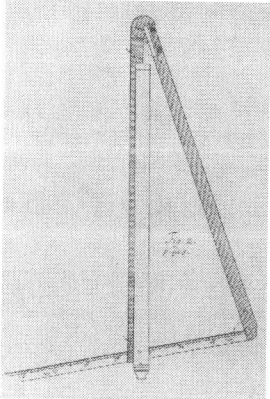

Figure 5.7 Hooke's reflecting instrument for measuring angles at sea, from his *Posthumous Works* (1705).

Instrument number 5, a portable quadrant, was probably also designed by Hooke for use at sea. Here the instrument seems to combine a plumb-line (probably partly encased, and incorporated into a handle) on the left and a telescopic sight on the right, pivoted at the centre of the quadrant arc. The function of the piece added below the sight is not clear, but it may have something to do with reading the scale. The latter is on a wide band at the limb, suggesting a diagonal scale, that is, one with 'diagonal' or 'transversal' lines drawn across arcs concentric with the primary scale and between its divisions, so as to effect a more accurate subdivision than would be possible on the primary scale itself.

The use of a plumb-line at sea would be to act as an artificial horizon, so that altitudes could be taken when the horizon is obscured. On 21 June 1665, Sir Robert Moray gave a meeting of the society some account of an altitude quadrant that Prince Rupert 'had contrived and made use of',

Figure 5.8 Instrument number 5.

which acted 'by a perpendicular'.[31] In this case the user could react instantly to having the target in his sight, by releasing a 'tricker' (a trigger) whereby the perpendicular would be 'clapt fast to the side of the quadrant' and give the altitude. This was thought appropriate for shipboard use, as the instant capture of the reading would mitigate the effect of the ship's motion. It also indicates interest in the plumb-line as an artificial horizon even for portable instruments. It cannot, however, be the instrument in the print, as in such a quadrant the sights would be on the upper radius and the perpendicular arm would give the reading.

Waller's edited version of Hooke's 'Lectures concerning Navigation and Astronomy', included in the *Posthumous Works*, incorporates an imperfect account of a sea quadrant using a perpendicular. These lectures date from the early 1690s, but Hooke often says he is drawing on inventions of some thirty years before (though he does not specifically say that of the quadrant). He makes much of the advantages offered by the design, because the perpendicular removes any reliance on a good horizon, and the lecture was illustrated by 'a Module', 'which I have here made . . . in order to make it the more intelligible'. Unfortunately Waller tells us that there was no surviving description or drawing, presumably because in the lecture Hooke explained the instrument using his 'Module'. However Waller had found 'a Rude Model of such an Instrument', which he described and drew himself.[32]

So, we cannot be sure that Waller's drawing (see Fig. 5.9) is of the model Hooke brought to his lecture, and it is not the instrument in the frontispiece.

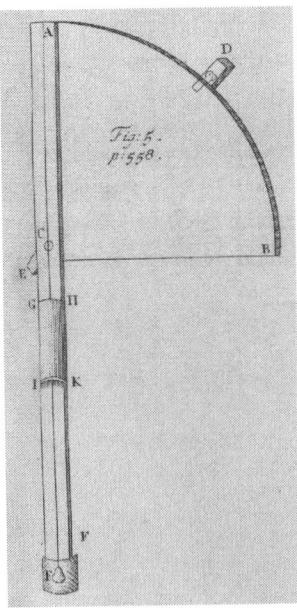

Figure 5.9 Sea quadrant designed by Hooke, from his *Posthumous Works* (1705).

The plumb-line is accommodated to the handle and Waller says the scale division is by diagonals, but the quadrant is set in an inverse orientation on the perpendicular. This allows the observer to see both the plumb-line and the target object together at the centre of the quadrant arc. There a diagonal mirror reflects the image of the target into the eyepiece, which is at right angles to the plane of the quadrant, and the observer, looking from the side, sees both the target and the plumb-line – a typical move of Hookean ingenuity.

While this is not the arrangement in the frontispiece, it is tempting to think that there the addition below the telescope, and perhaps the odd shapes around the object end of the telescope and the top of the perpendicular, are other modifications to facilitate achieving a reading. In Hooke's own list of the advantages of his instrument (before Waller takes over the description), he says that 'the Eye-glass' greatly enhances the accuracy of reading the scale, though nothing of the sort is indicated by Waller's drawing.[33] The addition below the telescope in the frontispiece, although it does not seem to be an eye-glass, may have a role in taking the reading.

Instrument number 13 coincides in some essential features with one Hooke showed to the society on 21 June 1665:

> Mr. Hooke produced a sextant contrived by himself, and explained the use and structure thereof, viz. that it was made after the manner of a pair of dividing compasses, there being two three-feet tubes opening upon a joint in the manner of the legs of compasses, and a long strait screw moving in two motions, serving to take angles very exactly.[34]

The compass hinge is clearly shown in the frontispiece, where we also have two equal and similar arms, which by this account are telescopic sights.

Figure 5.10 Instrument number 13.

Figure 5.11 Instrument number 17.

Unfortunately any connection between them is obscured by Fame's wing, though on the left we do see the beginning of some appliance for managing the tubes and quite possibly their connection and the instrument's measuring function. Further, this identification fits the emerging pattern, the 'sextant' being of an appropriate date and being a real object, since another Fellow, Sir Paul Neile, immediately proposed that it should be tried out. The repository later contained '2 square wooden tubes moving on a hinge'.[35]

Finally in this group, instrument 17 is a portable quadrant with micrometer subdivision. It has not been accurately depicted, as the pivot of the right-hand arm is not shown at the centre of the quadrant arc, which it must be for the instrument to function. Nonetheless, some features that characterise the originality of the design *are* clearly shown. It is evident that, as this arm, which is a telescopic sight, is moved around the limb, it carries with it a divided circle with an index that is turned in some way by this motion, so that the main divisions on the quadrant limb are subdivided on the circle. We have two descriptions of instruments of this type designed by Hooke, one from Thomas Birch's transcriptions of the minutes of the society's meetings in his *History of the Royal Society* (1756–7), the other contained in the 'Lectures concerning Navigation and Astronomy' published in the *Posthumous Works*, edited by Waller, who says he was unable to assign them a date.

An initial assumption might be that the motion of the index was by rack and pinion, but the accounts we have say that a 'roller' was moving on the limb. The record in Birch is worth noting in detail. On 22 February 1665, Hooke

> produced a new small quadrant contrived by himself, to make, by the means thereof, both celestial and terrestrial observations with more exactness than by the largest instruments, that had been hitherto publicly known. This quadrant was only of 17 inches radius, being by the contrivance of a small roller, that moved upon the limb of it, made so

accurate, that each degree was actually distinguished into 60 minutes, each of which minutes being about one third of an inch long, was actually divided into six parts, denoting every 10 seconds in a minute. The sights were likewise so contrived, though but short, as to be no less curious in distinguishing the parts of a minute in the visible object.[36]

This is surely the authentic voice of Robert Hooke, emphasising, as he does repeatedly in the *Animadversions*, that it is pointless to divide an arc into smaller angles than can be resolved by the eye. The record goes on to say that, although the plumb-line or 'perpendicular' was not much more than three feet long, 'by means of an index' its sensitivity was equivalent to a line of sixty feet. In the frontispiece the left-hand arm is blank, with no attempt to represent this 'perpendicular'; we know from the reference to 'the sights' (plural) that this too was a telescope.

Much of this coincides with what is probably the later description in *Posthumous Works* of a quadrant of a similar overall size (see Fig. 5.12).[37] We learn much more, however (more than would be appropriate to repeat here), from a detailed account given there of the 'index', mentioned by Hooke in 1665, for enhancing the sensitivity of the short plumb-line. The

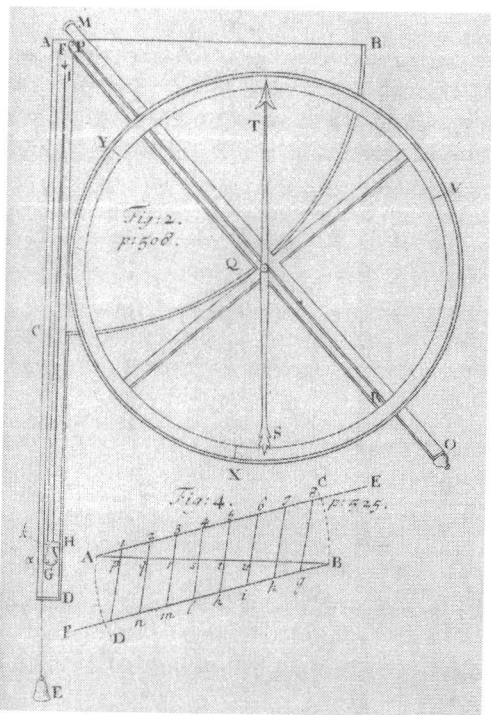

Figure 5.12 Hooke's micrometer quadrant, from his *Posthumous Works* (1705).

perpendicular was mounted on the outside of a square tube that accommo-dated a telescope. We learn also that a string is stretched around the brass limb of the quadrant, which is at least an inch thick, and passes around the 'roller' mentioned in the minute above, the roller also being made of brass, of a similar depth to the brass limb.

One obvious difference between the two accounts is the size of the circular scale compared with the radius of the quadrant. In *Posthumous Works* the circle is much larger, as Hooke tries to achieve an even finer subdivision. The propor-tionality of the quadrant in the frontispiece is more consistent with Hooke's account at the meeting of the society in February 1665. That account, however, mentions a 17-inch quadrant and a 3-foot plumb-line, proportions not reflected in the print, which is probably a further inaccuracy in the etched image.

An instrument of this type was preserved in the repository, an inventory of c.1730 listing:

> 2 wooden square Tubes of Telescopes turning on an axis, with a Quad-rant fasten'd to one & a brass circle to the other.[38]

This is probably Hooke's instrument of February 1665, as the illustration in *Posthumous Works* shows the moveable telescope as a round tube. It was a remarkable attempt to bring together a number of Hooke's ideas in a small instrument. The combination of a fixed telescope and a specialised plumb-line in one arm aimed to make the micrometric measurements of the movement of the other as versatile as possible for, as Hooke is recorded as saying, 'both celestial and terrestrial observations'. It was just the kind of design he had in mind when animadverting so passionately on the instru-ments of Hevelius.

An unusual longcase clock

Instrument number 4 has the general appearance of a clock in the style of Ahasuerus Fromanteel, who introduced the longcase pendulum clock into England and whose work interested several early Fellows of the Royal Soci-ety. Evelyn visited him in May 1661 with Huygens, 'to see some *Pendules*'.[39] A pendulum clock was a novelty in itself but, if this is a clock, it is a very unusual one. The most obvious peculiarity is that it is square in plan and the hood is surmounted by a tall pyramid.

If the pyramid is functional, rather than decorative or symbolic, the most natural horological explanation is that the clock is controlled by a conical pendulum, that is, one where the motion of the bob describes a circle. On 13 June 1666, 'Mr. Hooke exhibited a new contrivance of a circular pendulum applicable to a watch, and moving without any noise, and in continued and even motion without any jerks.'[40] Evelyn was present and noted in his *Diary*: 'To the R: Society, where was brought the new Pendulum'.[41] This 'watch' was certainly a clock, as noted in the society's minutes of 18 July, when already Brouncker was able to affirm that he had conducted a successful trial over

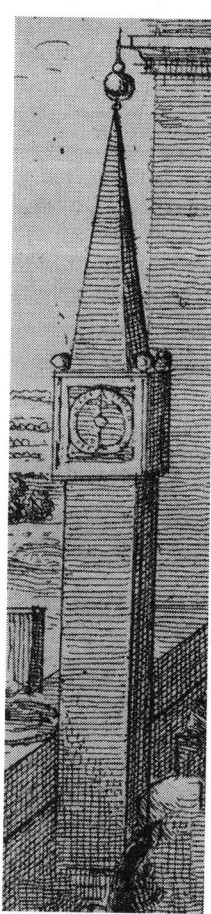

Figure 5.13 Instrument number 4.

four days.[42] In a further development in January and February 1667, there was sustained discussion around Hooke's 'new clockwork' to improve the timekeeping of his circular pendulum by rendering its motion isochronous.[43] Hooke 'produced a circular pendulum so contrived' on 21 February.[44]

This must have seemed a potentially outstanding development in horology, since the clock would deliver, as Hooke pointed out at its first presentation, a smooth motion, unlike mechanical timepieces up till then. Sprat selected it for mention in his *History*: 'A new kind of *Pendulum Clock*, wherein the *Pendulum* moves circularly, going with the most simple, and natural motion, moving very equally, and making no kind of noise'.[45] The smooth motion made it suitable for regulating drives for equatorial instruments in astronomy. Hooke included it in the earliest design for such an instrument. Although his quadrant was never built, his illustration in *Animadversions* (see Fig. 5.14)

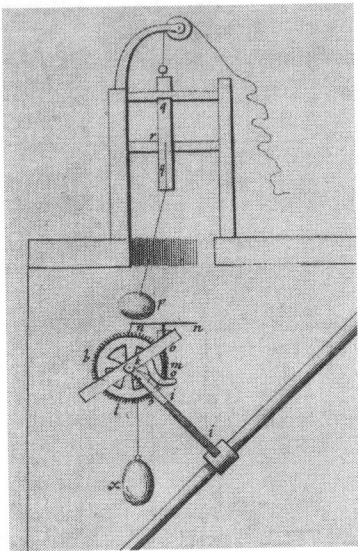

Figure 5.14 The clockwork drive for Hooke's proposed equatorial quadrant, controlled by a conical pendulum, shown in a detail from fig. 2 in plate 2 of Hooke's *Animadversions On the first part of the Machina Cœlestis of . . . Johannes Hevelius* (London, 1674).

shows the pendulum mounted above the movement of the clock drive. Here he says that he had invented the circular pendulum 'long since, in the year 65', had 'brought it into use in the year 1665', and communicated it to the Royal Society in 1666, 'both as to the Theory and Practick thereof'.[46] The theoretical component would have included the explanation of isochronous motion achieved by confining the bob to a parabola of revolution.

Hooke's concern over dates was a reaction to Huygens's publication of such a design in *Horologium Oscillatorium* (1673). In conveying these concerns to Huygens, Oldenburg affirmed that some years ago Hooke had shown the properties of such a pendulum to the society and even had some clocks made that were seen by a number of foreigners.[47]

The dial visible in the print has a single hand, which may seem problematic for a clock, but the same might be said of the longitude timepiece, no. 1 above, which definitely is a clock. In the present instance, however, the hood is square and it would be perfectly appropriate, especially in so experimental a clock, to present different indications on its several faces.

What alternative identities are there for this instrument? Two might be offered for consideration. Wren and then Hooke spent time on evolving designs for a 'weather clock', that is, a self-recording device for several registers of the weather, all maintained by a single clockwork drive. The clock elements in the two illustrations we have of Wren's work show the stylistic influence of Fromanteel (see Figs. 5.15, 5.16). This development would have

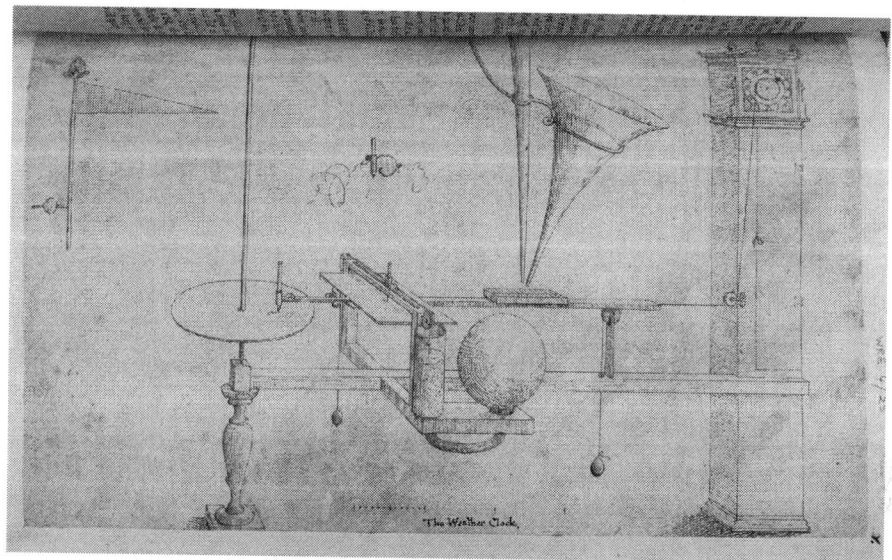

Figure 5.15 Wren's drawing of a weather clock, from the 'heirloom' copy of *Parentalia: or, Memoirs of the Family of the Wrens*, edited by his son in 1750, now at the Royal Institute of British Architects.

Figure 5.16 Wren's design for a weather clock in a drawing presented to the Royal Society in 1663, from the society's register book.

been at least as exciting as a conical pendulum clock (indeed the two might even have seemed complementary, since a uniform drive would be appropriate for continuous recording), but the narrative of the weather clock is not so good a fit with our growing understanding of the ambition and timetable of the frontispiece.

The first account we have of a working weather clock comes from the visit the French traveller Balthazar de Monconys made to Wren at All Souls College, Oxford, in June 1663. He was shown a weather clock and his description coincides fairly well with a manuscript drawing by Wren (see Fig. 5.15) preserved in the volume known as the 'Heirloom' copy of *Parentalia* (the account of Wren's life and work edited by his son) at the Royal Institute of British Architects.[48] There is no clear link to the Royal Society at this stage, but Wren submitted a rather different description and drawing in December 1663 (see Fig. 5.16).[49] There was then discussion of a way, expense being one of the challenges, 'of reducing this engine into practice'. There is no evidence that that was ever done.

We should note that in both of Wren's designs the clock is shown with a single hand. In the earlier example, this is because the motion is taken directly from the spindle for the hour hand, this being an adaption of an existing clock. The hour hand has been replaced by a pulley wheel and an index, which registers hours on the dial. The later illustration is of a proposed design rather than an existing instrument, and Wren is making it evident, with dotted lines indicating the hidden rackwork, that the motions for both registers are taken from the hour motion of the clock. The single hand of the clock registers time, not any indication of the weather.

If we consider that the clock in the frontispiece does not look like the one in Wren's first drawing, that there is no suggestion that this was ever shown at or explained to the Royal Society, that by late 1663 Wren had moved to what would have been a more robust design, which is definitely not the one in the print, and that this itself was never made, the case for this representing the weather clock looks distinctly weaker than that for Hooke's conical pendulum clock, which is from the right period and was made.

In September 1664 Hooke was ordered to 'contrive' a weather clock 'as well and as cheap as he could, for the use of the society', but after many reminders he showed it 'now finished' only in May 1679.[50] In 1681 Nehemiah Grew, in his catalogue of the society's repository, *Musæum Regalis Societatis*, described a clock begun by Wren and augmented by Hooke, with no fewer than six registers of the weather.[51] A mechanical weather clock of this sort will register and record outside the clock case and movement, making its function evident. It was certainly an ingenious development for the society but, were it included in the frontispiece, it would be odd to have omitted or disguised, or at least minimised, any indication of its purpose.

The second alternative to Hooke's conical pendulum clock is that this is the longcase clock by Fromanteel that Bishop Seth Ward presented to the society to commemorate Lawrence Rooke, a prominent early Fellow, who died in 1662.[52] This is recorded by William Derham in his account of pendulum clocks in his *Artificial Clock-Maker* (1696), where he wrote: 'One of the first Pieces that was made in *England*, is now in *Gresham-Colledg*, given to that Honorable Society by the late eminent *Seth*, Lord Bishop of *Salisbury*: which is made exactly according to Mr *Zulichem*'s Directions'.[53] It was still at the Royal Society in 1756, when Thomas Birch transcribed the inscription on it recording Ward's gift in his *History of the Royal Society*.[54] There is no record of that clock being square or having a pyramid above the hood, and that identification would not account for such a form, unless we might think of it as some kind of memorial obelisk. More importantly perhaps, there is no suggestion of that clock playing a part in the experimental or mathematical work of the society.

A long telescope

Instrument number 6 is the 60-foot telescope on which Hooke worked with the instrument maker Richard Reeve in 1666–7.[55] Hooke provided Oldenburg with details of the telescope and a sketch of the mounting, in a letter of around 20 February 1667, with the intention that an account would be passed to Johannes Hevelius in Danzig.[56] It seems that Oldenburg wanted to send a fuller explanation, for he has added to the letter information that can only have come from Hooke, as well as a number of letters to key Hooke's text to the sketch with greater clarity.[57] However Oldenburg made a number of mistakes in applying this key and, perhaps because he realised his confusion, Hooke supplied a more detailed drawing with all the key letters placed correctly. This was sent to Hevelius with Oldenburg's letter of 27 February and has been preserved in the Bibliothèque Nationale, Paris (see Fig. 5.18), although the key that must have accompanied it (since Oldenburg did not attempt in his letter to elucidate the letters on the drawing) seems not to have survived.[58] The image of the telescope in the Sprat frontispiece is very close to the Hooke drawing, down even to the tiny observer, with his cloak and hat. The coincidence of detail and the secure date for the drawing not only throw important light on the making of the frontispiece but also provide a useful *terminus post quem* for its execution.

The fact that the frontispiece includes the 60-foot telescope is further evidence that the aim was to illustrate the most up-to-date instrumentation in the society's repertoire, especially where it represented recent and

Figure 5.17 Instrument number 6.

current work. The 60-foot telescope by Reeve had displaced a 36-foot refractor by the same maker, mounted on a mast and pulley at Gresham College, and also drawn by Hooke (see Fig. 5.19).[59] In fact the mast had been set up in 1658 by Wren and Sir Paul Neile for what Wren refers to as a 35-foot telescope, while Hooke says that Wren also used a 28-foot telescope there.[60] A new tube, provided by the Royal Society, was mounted for Hooke's use in September 1664, and contained, by Hooke's account, a 36-foot object-glass. The subsequent tube (the third, at least) to be suspended there was for the 60-foot lens, made by Reeve and purchased by Robert Boyle, the first observations by Hooke being recorded from June 1666.[61] It is in line with the overall thrust of the frontispiece that the telescope depicted is the latest and technically most demanding of the series. It had taken the society a few years to fund the purchase, relying in the end on Boyle's generosity, and it must have been considered an important and recent coup that they had access to what was understood to be the finest telescope in the land.

We can be sure that the Gresham mast survived through these changes of telescope, because it had some very distinctive features. Its depiction in the frontispiece is not, of course, complete in every detail: the essential winch for raising and lowering the tube is absent, though it is included in Hooke's drawing. Yet details are present at the top of the mast that secure its identity. Wren made several models of his solution to the Saturn problem, prior to the publication of Huygens's hypothesis of a detached ring, and one of them was mounted on this mast, as he wrote to Sir Paul Neile in 1661:

> The hypothesis made more durable in metal was posed on the top of that Obeliske, which was erected at Gresham Colledge in May 1658 . . . to rayse the 35 foot Telescope.[62]

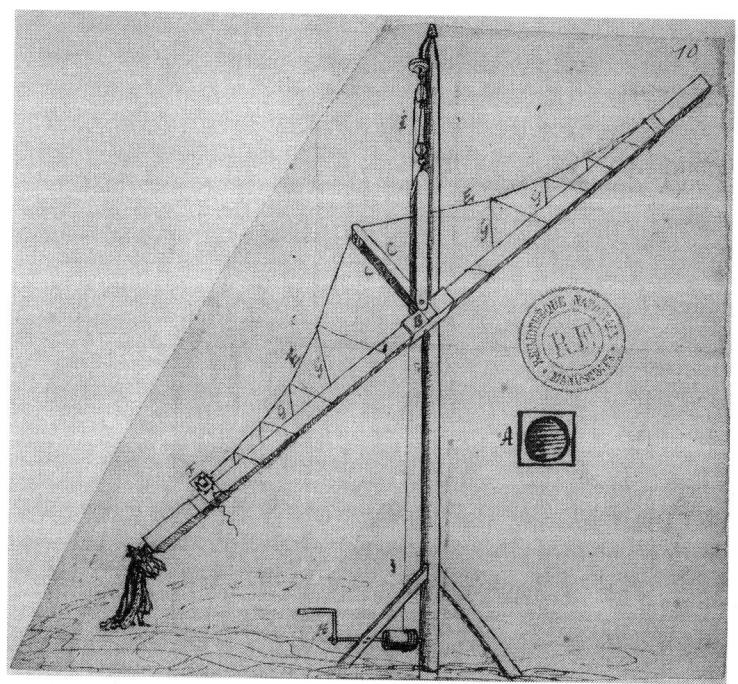

Figure 5.18 Hooke's drawing of the 60-foot telescope sent by Oldenburg to Hevelius in February 1667.

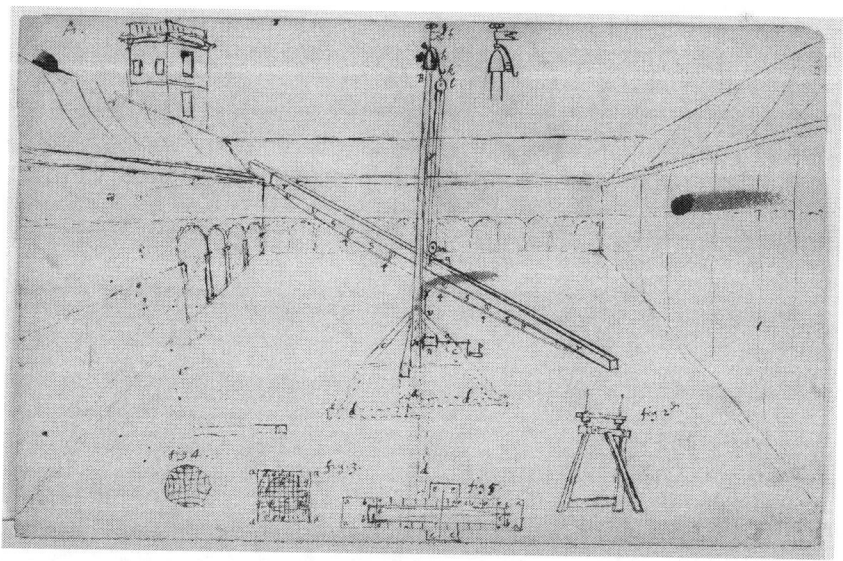

Figure 5.19 The 36-foot telescope erected on its mast in the Great Court at Gresham College: drawing by Hooke preserved among his papers at the Royal Society.

This is drawn twice – somewhat like a ring in the main depiction and three little circles in the detail – set above the pennant in Hooke's drawing of what he refers to as the 36-foot telescope: the identification of the model is confirmed by Hooke's lengthy description accompanying his drawing.[63] Viewing the model at a distance from different positions with another telescope would test how closely it could present the telescopic appearance of Saturn. The sketch of the 60-foot telescope Hooke initially sent to Oldenburg is rather diagrammatic and does not include this kind of detail, and the redrawn version sent to Hevelius has subsequently been trimmed, so that the top part of the mast has been lost. However both the pennant and even Wren's model of Saturn are present in the frontispiece.

Comparing Hooke's more detailed drawing and the print gives some indication of the accuracy of the latter and the care taken over its production. An obvious discrepancy is the absence of the winch in the print. Further the central, orthogonal support for the rigged bracing of the tube is either missing or impossibly placed on the other side of the mast from the tube. On the other hand, the rigging itself is accurately rendered and even the tiny detail of Wren's model for Saturn is present. If we generalise from this image alone, which may not be wise, it would seem that not every feature, not even all the prominent ones, are included, but those that are reflect features that were present.

Chemistry and pneumatics

Apparatus grouped together as number 7 consists of a chemical furnace (the markings on it representing the vents and fire-hole) and a collection of chemical glassware. Among the glass apparatus is a distillation train with a sinuous column and a 'head' (with a supporting stand) at the top for carrying condensed vapours to a receiver flask at the bottom. The long beak is a typical form used to improve the efficiency of distillation, especially of more volatile liquids like alcohol-based materials for medical cordials. A sinuous tube of this type appears in Sebastien Leclerc's etching of Louis XIV's visit to the Académie des Sciences (Fig. 1.2), and other chemical equipment similar to that shown here appears in his vignette in the *Histoire des Plantes* (1676) (Fig. 5.21).[64] There are also a retort feeding a receiving vessel and a second distillation train comprising a flask (cucurbit) with an alembic fitted to it, that is, a dome to collect the vapours from the substance in the cucurbit with a tube leading to a globular receiver.[65]

As early as December 1662 Hooke was asked to perform a distillation, one of a number of early references to chemical operations, but on 8 June 1664 something more considered was attempted: Dr Jonathan Goddard was appointed to chair 'the chemical committee', to meet twice a month at his

Figure 5.20 Apparatus grouped together as number 7.

Figure 5.21 Sébastien Leclerc's etched vignette showing the chemical equipment of the Académie des Sciences in Paris, from *Mémoires pour Servir à l'Histoire des Plantes* (1676).

lodgings, which were in Gresham College, where he was Professor of Physic.[66] Oldenburg wrote to Boyle on 14 February 1666, 'We are now undertaking severall good things, as the Collecting a Repository, *the setting up a Chymicall Laboratory*, a Mechanicall operatory, an Astronomicall Observatory, and an Optick Chamber'.[67] Of these, the repository became a reality and the record of the frontispiece suggests that at least some equipment was assembled for the laboratory.

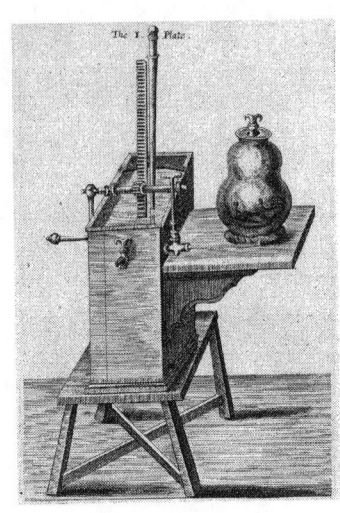

Figure 5.22 Instrument number 8.

Figure 5.23 The version of Boyle's improved air-pump depicted in his *Continuation of New Experiments* (1669).

Instrument number 8 is an air-pump following Boyle's second published design. While this was perhaps the most challenging and influential instrument associated with the society in its early years, so an expected inclusion in the frontispiece, what we see is not the familiar, iconic image of Boyle's pump. In his *Continuation of New Experiments Physico-Mechanical*, published in 1668, though dated 1669, Boyle illustrated an air-pump (Fig. 5.23) that was significantly different from the one described and depicted in his *New Experiments Physico-Mechanical* of 1660, having a detachable receiver and a barrel submerged in water, to keep the sucker wet and airtight.[68] He had been using a redesigned pump in Oxford since early 1662, having given his first pump to the Royal Society the previous year. At the society also work began on a new pump in 1662, promoted by the appointment of Hooke as Curator of Experiments in November, and by early 1663 the society had a working pump on the same principles as that which Boyle was using in Oxford. As well as being in communication with Hooke, Boyle himself was often in London. In their book, *Leviathan and the Air-Pump*, as well as explaining these parallel developments, Steven Shapin and Simon Schaffer stress that pumps were frequently modified and adjusted over this and the following period.[69]

These developments explain why the air-pump in the frontispiece looks both like and unlike Boyle's illustration of 1668. The two pumps had had parallel developments in different but closely related contexts. It is clear from the minutes of their meetings that for several years prior to the publication of

Sprat's *History*, the society had immediate access to a pump capable of performing experiments facilitated by the great advantage of the new design.[70] This was the provision of a flat plate, capable of supporting receivers of different sizes and shapes, where experiments could readily be set up before working the pump, which then evacuated the receiver by a pipe opening through the plate.

The most obvious discrepancies between the frontispiece and Boyle's illustration are the different receiver (not surprising since interchangeability of receivers was an advantage of the new design) and the difference in the stand and the support for the water tank containing the submerged barrel. More significant perhaps is the different orientation of the handle and therefore the rackwork of the piston rod, and the absence in the frontispiece of the top of the long rod that served as a valve, passing through an aperture in the 'sucker' or piston. This rod was detachable, so its absence is understandable.

This instrument neatly supports the growing impression that the instruments in the frontispiece are not thoughtless selections decorating the space or chosen simply to impress. They are real objects that both represent the recent and current activities of the society and were physically present in the society's world. The distinction between the London and Oxford pumps has allowed us to make these points with even greater precision.

Thermometers

Instrument number 11 seems to be a sealed spirit thermometer, its long stem given a serpentine form, so as to combine a longer range, consistent with the large reservoir at the bottom, with a manageable size. No instrument of this form survives, but such thermometers would have been easily

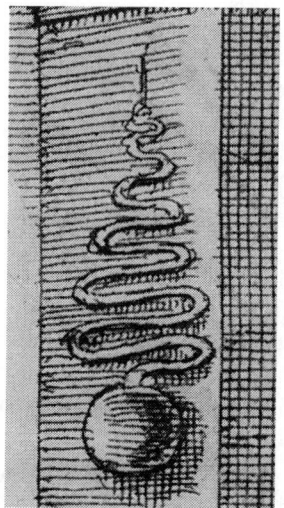

Figure 5.24 Instrument number 11.

Figure 5.25 Spiral thermometers made for the Accademia del Cimento, surviving in the Museo Galileo, Florence, inv. 193, 194/a, 194/b.

broken in use and not preserved. The Accademia del Cimento in Florence had spirit (that is, alcohol) thermometers in the mid-seventeenth century with large reservoirs and long helical stems (Fig. 5.25). This form and others were illustrated in the *Saggi di Naturali Esperienze*, prepared by the Secretary of the Academy, Lorenzo Magalotti, and published in 1667.[71] A number of these thermometers do survive and those with helical stems in particular clearly depended for their production on the outstanding skills of specialist glassworkers. While the capabilities of English glassmakers were advancing in this period,[72] this type of helix would have been a singular challenge.

Evelyn gives a 'thermoscope' a roughly helical glass stem in an illustration (Fig. 5.26) in his unpublished *Elysium Britannicum* (the draft, which had later additions, was prepared between c.1657 and the mid-1660s); the stem is drawn alternately in front of and behind a supporting vertical stave.[73] This 'brittle plant', as he calls the instrument, was intended for the garden, to inform the gardener of the 'disposition' of the air. The stem connected an airtight vessel at the foot, which was a reservoir for coloured water, to a bulb at the top. It was not a thermometer, but a weather-glass, having an aperture at the top and responding to pressure as well as temperature. A marginal note to the *Elysium Britannicum* manuscript, which is in the British Library, has the reference 'Schotti: 231', and the German Jesuit natural philosopher, Gaspar Schott, describes just such an instrument, his 'Thermoscopium Æstivum' or 'summer thermoscope', on p. 231 of his *Mechanica Hydraulico-Pneumatica* of 1657.[74] Evelyn's device, however, seems not to be merely an idea copied from Schott, but a real instrument:

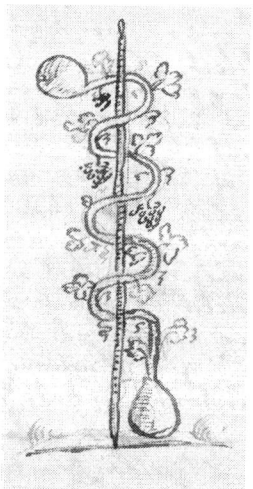

Figure 5.26 Evelyn's drawing of a 'thermoscope', from his *Elysium Britannicum*.

'We have severall formes of these Weather-glasses, placed artificially in Rock-worke.'[75]

It seems very likely that Evelyn had sourced his stems from English glass-works. Their form was not unlike the 'sinuous column' that is part of one of the distillation trains in the chemical glassware described above, and both these illustrations probably constitute evidence for what was at least begin-ning to be produced by English glassworkers. The same is probably true of the tapering serpentine stem of instrument 11, that is, that it too is likely to have been of English manufacture. It is clear from the manuscript that Evelyn's drawing is part of the original draft of *Elysium Britannicum* (c.1657 to mid-1660s), but he has added a note: 'Improve this by the R[oyal] Society way.'[76] While we cannot say when the addition was made, it might well be a reference to the innovation of the sealed thermometer, which would bring his device much closer to the instrument in the frontispiece. As to the form of instrument 11, a helical thermometer would have been even more awkward to hang on a wall in the manner shown than the serpentine shape it is given in the print. Although the latter form is not otherwise known at this time, it is more likely that the etching is a fair representation of the true shape of the stem than a very poor one of a Italian-style helix.

A Florentine thermometer does, however, come into the story. In his *New Experiments and Observations Touching Cold* of 1665, Boyle reports that his own attempts to make thermometers were helped by his being shown one brought from Florence:

I found my work much facilitated by the sight of a small seal'd Weather-glass, newly brought by an Ingenious Traveller from *Florence*, where it seems some of the Eminent *Virtuosi*, that enobled that fair City, had got the start of us in reducing seal'd Glasses into a convenient shape for Thermoscopes.[77]

According to a later account by the astronomer Edmond Halley, the traveller was the diplomat Sir Robert Southwell;[78] this is entirely plausible, as Southwell was known to Boyle and in 1660 spent three months in Florence, where he 'frequented the circles of virtuosi', returning to England in 1661.[79] It is unlikely that the thermometer he brought home was of the helical form, not least because Boyle says it was 'small', but was probably one of the more common fifty-degree thermometers, which were indeed small and had straight stems (Fig. 5.27).[80] Tiny glass beads applied along the outside of the stem marked the degrees – white beads, with every tenth one black. These 'degree' scales did not conform to any standard.

Discussions at the society help us to date Boyle's work and the growing English interest in thermometers of this type. As early as December 1662, it was proposed that 'a sealed up thermometer' should be sent to record 'the degree of cold' in Iceland;[81] Hooke distinguishes this instrument from a 'common weatherglass'.[82] On 3 September 1663, John Wilkins urged the Fellows to apply themselves more purposefully to the existing project of making a 'history of the weather', one step he recommended being to ask Wren for 'a scheme for his weather-engine, formerly proposed';[83] this resulted in their receiving Wren's developed design in December, as was mentioned above. On 9 September Boyle proposed a series of observations specifically with 'a sealed thermometer with spirit of wine'.[84] Such instruments could be sent to distant observers.[85] Wilkins was interested to know the origin of 'this kind of thermometers with spirit of wine, sealed up',

to which Mr Boyle answered, that they were brought out of Italy; and that he thought, that he had received thence the first thereof here in England.[86]

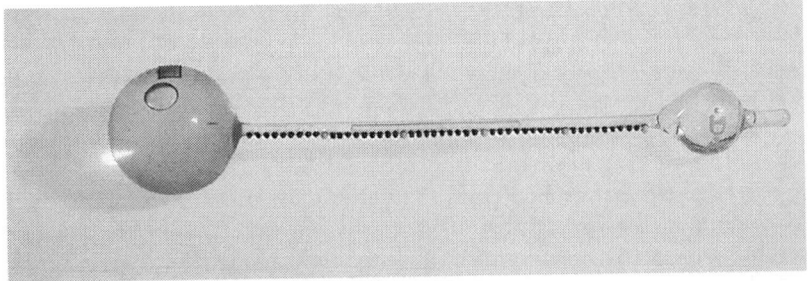

Figure 5.27 A 'fifty-degree' thermometer, with a straight stem, made for the Accademia del Cimento, surviving in the Museo Galileo, Florence, inv. 173.

Boyle's *New Experiments* and Hooke's *Micrographia*, both published in 1665, explain that sealed thermometers had come to be made in England, Boyle having found 'a person eminently dexterous in making such Instruments' and Hooke's trials being 'at last brought to a great certainty and tenderness'.[87] Although Hooke refers to stems of up to four feet in length, both authors assume a straight form. Instrument number 11 was probably an unusual example and number 18, another sealed thermometer but with a straight stem, much more typical. This example has both a bulb for the spirit reservoir and a smaller one at the top, features Hooke called the 'body' and 'head' respectively in 1665,[88] the year Boyle illustrated an instrument of this general type (see Fig. 5.29).

By the mid-1660s sealed thermometers had become fairly common instruments at the Royal Society and were in regular use. In February 1664 it was ordered that 'a couple of thermometers should be made ready . . . and carefully put up by the operator' to be sent to a correspondent in Iceland.[89] In February 1665 'a *common* sealed weather-glass' (our emphasis; it is referred to in the same record as a 'thermometer') was tried in the condensing engine; it was not affected by the compression of the air.[90] In October 1667 the society's 'operator' (not named, but Richard Shortgrave) could be entrusted with making a thermometer for the Queen.[91]

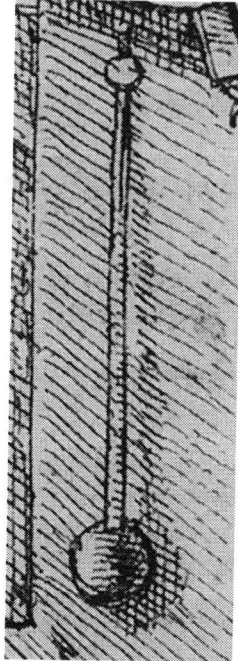

Figure 5.28 Instrument number 18.

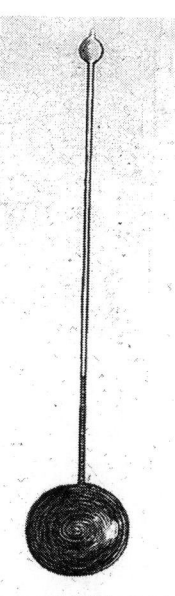

Figure 5.29 Thermometer, fig. 3 from plate 1 of Boyle's *New Experiments and Observations Touching Cold* (1665).

A sea-sounder, perhaps

Sounding the depth at sea without a line was a sustained interest of the society from an early date. It headed a list of 'experiments' recommended to the newly elected Fellow Edward Mountagu, Earl of Sandwich, in June 1661, on his voyage to Lisbon in command of an English fleet, whose mission included negotiating the King's marriage to Catherine of Braganza.[92] At this stage the proposed apparatus was fairly crude. A wooden float would be dragged down to the seabed hooked to a lead in the shape of a '7'. This might be expected to fall over and release the float, whose return to the surface would be timed. Trials in known depths would establish a relationship for use elsewhere. While Sandwich had many other things on his mind, before he returned in May 1662 with the infanta, soon to be Queen of England, he did make a report, though not about soundings.[93]

The society developed their apparatus and arranged trials at sea.[94] By March 1662 the wooden float was being specified as a ball and by September 1663, when Hooke described his 'new' design, the weight also was drawn as a sphere.[95] Hooke's main contribution was a more reliable arrangement for releasing the float. The connection between the weight and float was now secured by a spring, which was released by the continued brief motion of the float, after the weight had reached the bottom first. This interest remained active at the society and it is noteworthy that the new sounder was thought appropriate to show the King on his much-anticipated visit, which never came to pass.[96]

In January 1666 Oldenburg published in the *Philosophical Transactions* the directions for seamen 'bound for far Voyages', that had formerly been compiled by Laurence Rooke.[97] No doubt at Hooke's instigation, an 'Appendix' appeared in the following issue with his descriptions and illustrations of his sounder and his device for sampling the sea at depth (Fig. 5.31).[98] The directions were expanded and printed as a separate tract, with copies

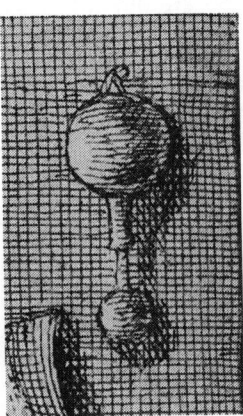

Figure 5.30 Instrument number 12.

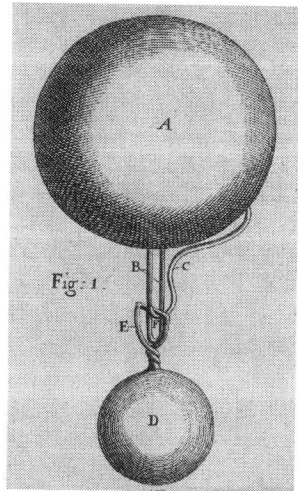

Figure 5.31 Detail from the plate in *Philosophical Transactions*, no. 9, showing the depth sounder.

deposited at Trinity House to be issued to seamen, and this expanded tract occupied a whole issue of *Philosophical Transactions*, number 24, published in April 1667.[99] Here a new plate had re-engraved copies of the sounder and the sampler, together with other instruments.

In the evolution of sounder designs, it is the virtually identical illustrations in the two issues of *Philosophical Transactions* that come closest to instrument 12 in the frontispiece. In Hooke's earlier drawings, from 1663, the float and the weight below are of a similar size, but in early 1665, the float is significantly larger and the proportion between the two spheres is roughly the same as in the frontispiece. It seems significant that the design had reached this stage in the core period for many of the instruments in the print. It is also worth noting that the sounder had actually been made and tried.

Thus far, the history of the society's interest in sounding without a line seems promising for securing the identity of instrument 12, but there is a serious difficulty. The connection between the spheres in the frontispiece is presented clearly but it consists of two tubes, one simply fitting inside the other, and is nothing like Hooke's release mechanism. This may be considered sufficient to disallow instrument 12 as a sea-sounder but, until we have a stronger candidate, we might admit at least the possibility that the connection is not accurately rendered.

One barometer or two?

Hooke first explained the wheel barometer (instrument number 19) to the society on 30 December 1663, when he 'produced a little engine for making the descent of quicksilver in glass-canes more discernible'.[100] The

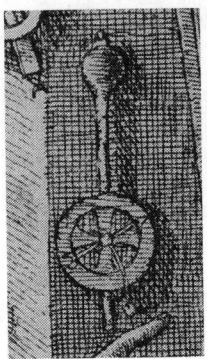

Figure 5.32 Instrument number 19.

idea was that a float on the open surface of the mercury would be attached to a line that would hang over a wheel, and connect to a counterweight to keep the wheel and line in contact. A light index attached to the wheel would move around a graduated circle, thus revealing, and magnifying, the changes in the level of the mercury. At this stage Hooke's 'little engine' had not been fitted to a mercury column and, although this was ordered to be done, there is no report in subsequent minutes of its successful application.[101] The earliest evidence of a complete, working instrument comes in a letter from Hooke to Boyle of 6 October 1664, where Hooke reports that, since moving into Gresham College five weeks previously, he had 'constantly observed the baroscopical index (the contrivance, I suppose, you may remember, which shews the small variations of the air)'.[102] So, we know of at least this example of a wheel barometer at Gresham during the core period represented by the instruments in the frontispiece. A figure and a detailed description were published in *Micrographia* (1665) (Fig. 4.8b).[103]

The instrument was also under active improvement. Hooke wrote to Boyle on 21 March 1666 that he had given Shortgrave directions for making Boyle 'a wheel baroscope . . . by a new way';[104] it was probably this that was described and illustrated in issue No. 13 of the *Philosophical Transactions* for 4 June 1666 as 'A New Contrivance of Wheel-Barometer, much more easy to be prepared, than that, which is described in the Micrography'.[105] The difference was in the arrangement of the open end of the barometer for receiving the float.

The accounts in *Micrographia* and *Philosophical Transactions* are both accompanied by figures and a drawing by Hooke survives at the Royal Society (Fig. 5.33).[106] However, the contemporary illustration where the proportion between the mercury column and the circular scale comes closest to the instrument in the frontispiece is in fact one in the body of Sprat's *History* (Fig. 5.34).[107] All four illustrations show what Hooke calls a 'bolthead' at

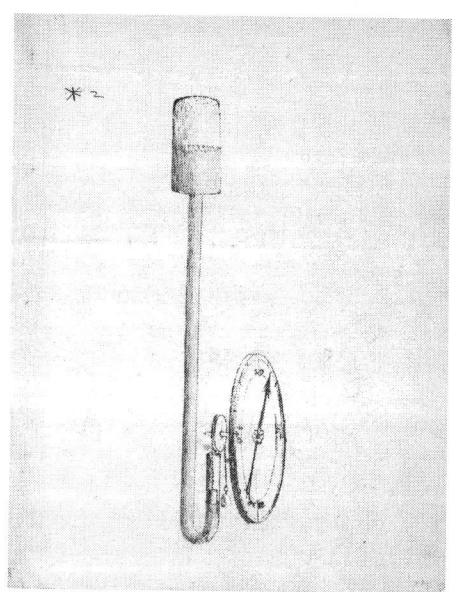

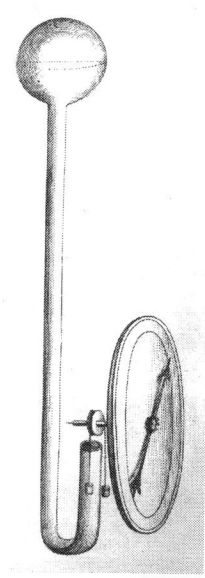

Figure 5.33 A drawing by Hooke of a version of his wheel barometer, unusual in that the 'bolthead' at the top is not spherical. Cl. P. 20, no. 32.

Figure 5.34 Detail from plate in Sprat's *History* showing a wheel barometer.

the top, though he says the device will work also with a simple tube; the bolthead is spherical, except for the manuscript drawing, where it is close to cylindrical. The figure in the body of Sprat's *History* is rather diagrammatic, but there the mercury tube does appear to extend below the scale, as it does, though to a greater extent, in the frontispiece, but not in the other two figures. The *Micrographia* and frontispiece figures have openwork graduated circles; the other two appear to be solid. An unusual feature of the instrument in the frontispiece is that the index has an additional arm extending at right angles from the pivot.

Instrument number 15 has the overall disposition of a wheel barometer, but an examination of its detail suggests that it is a different type of instrument. The similarities are a circular scale, with an index, and a tube rising vertically from behind. The tube terminates, however, not with a bolthead but a ferrule, from which emerges a thin vertical rod ending with a ball or knob, and to the left an oddly but very deliberately shaped attachment. The tube itself perhaps seems more regular and sturdy than that of the barometer, suggesting that it may be of metal, rather than glass. At its lower end is another ball or knob, and below this the short attachment to the disc is narrower than the tube above. The disc itself is solid. In the barometer Hollar indicates clearly that the wall is visible through an openwork circle, but here the disc has a different pattern of lines to distinguish it from the wall

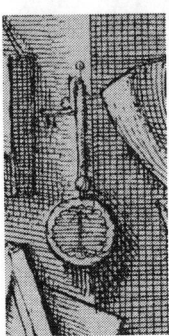

Figure 5.35 Instrument number 15.

behind, while, since there is an index, we may assume that the circular limb carries a scale.

We might use the index pivoted at the centre of the disc as a diagnostic tool for considering the identity of this instrument. An index arm might seem so obvious a feature in an instrument that we can take it for granted, but this is not so for the mid-1660s. It had an established role in clocks and watches, and in the related domain of mathematical instruments, particularly astronomical instruments and globes, but was a newcomer to instruments for registering the properties of things. In the latter cases, the indexes we know had light mechanical actions, attuned to the delicate movements they were designed to register. We have already encountered this in the wheel barometer, and there is little reason to imagine that the print would display a further example of such a device. Another instrument with a light index was Wren's application of a float and line to the weather-glass (an open air-thermometer, which responds to the combined effects of temperature and pressure), but that would imply a reservoir for air, which is not present.[108]

A third meteorological instrument of the period requiring a light index was the hygrometer, but Hooke's design was quite different from the instrument shown, the versions we know having, for example, a horizontal dial.[109] In October 1663 Jonathan Goddard, who, as mentioned, was resident in Gresham College, 'proposed for an hygroscope the contrivance of a lutes-tring with pullies and a cylinder'.[110] That sounds promising, and instruments of that type are known, but there is no recorded response to a reminder to Goddard to 'provide his hygroscope'.[111]

Might the index on instrument 15 be moved by a stronger mechanical action, such as was the case for the clocks in the print, having index arms of a generally similar type? A close look at what we might assume is the raised limb on the dial shows that its inner boundary is 'crinkled', which may be Hollar's attempt, at a very small scale, to show that there are inner teeth on an upraised band. It might even be the case that the lower end of the index engages with these teeth, but that is far from clearly shown. Any rackwork feature would imply some relatively strong mechanical action, such as a spring balance for weighing. For that specific suggestion, however, it is not clear how the movement of a spring, for example contained in the vertical cylinder, might be registered by the index.[112]

Two other instruments attracting interest at the time, and sitting on the 'stronger mechanical action' side of the fence, are the odometer, or way-wiser, and the pedometer. Both measured relatively long distances, the former by registering the revolutions of a wheel on a road, whether pushed along by hand or incorporated into a carriage, the latter by counting the steps of a steady walker, or even a horse.[113]

Consider first the odometer.[114] The dial of instrument 15 would then be a wheel, and the vertical cylinder part at least of the rod extending to the user's hand. The overall length of the instrument, to judge by the known dimensions of the micrometer quadrant nearby (not, admittedly, a reliable inference), might be three feet. The rod, however, could be capable of extension and the 'oddly shaped attachment' could be a handle.

Both Wren and Hooke had interests in sophisticated designs of the odometer – examples that did more than simply measure distance – though respectively too early and too late for the content of the print.[115] Wilkins showed his way-wiser for a coach at a meeting in November 1662 and was asked 'to leave his first engine of this kind with the society', implying that he had more than one.[116] The repository did indeed acquire a coach way-wiser from Wilkins.[117] This cannot be instrument 15, but it is worth noting that one of the instruments Wilkins showed Evelyn in Wadham College in 1654 was a 'Way-Wiser', and that Evelyn seems to distinguish this from a device he saw in 1657, namely 'the application of the Way-Wiser to a Coach'.[118]

If instrument 15 could be a pedometer, the 'oddly shaped attachment' might be for a saddle.[119] Perhaps it is possible to imagine something moving in the vertical tube that might click the index on at each stroke, the position of the index being retained between strokes because one end engages the 'crinkled' (i.e., toothed) inner edge of the limb on the dial. That, of course, is very speculative, and for now instrument 15 has no secure identification.

The gun

One of the most conspicuous objects in the print – Francis Bacon may even be pointing to it – is number 16, the firearm. Gunnery touched the society's interests at several points, including the experimental study of ballistics and of gunpowder, one of Bacon's three exemplars of Renaissance invention. The society discussed guns large and small – artillery and firearms – and they acquired two notable examples in the period before Sprat's *History*. One

Figure 5.36 Instrument number 16.

was another gift from Wilkins in November 1663, an airgun or 'wind-gun', which, being an impressive example of the 'spring of the air', was relevant to pneumatic as well as ballistic demonstration, and was preserved in the repository for many years.[120]

Later the same month Dudley Palmer, not a prominent Fellow but named as a member of Council in the first charter, presented the society with a 'very artificial gun of Caspar Calthof's contrivance' that could be loaded with seven bullets and fire them without reloading.[121] The Kalthoff dynasty of gunsmiths was famous for making the 'Kalthoff repeater', and two of their number, both called 'Caspar', probably father and son, spent significant periods working in England.[122] Either could have been the maker of this gun. The younger Caspar Kalthoff had been an instrument maker in Dordrecht and in 1655 had provided the Huygens brothers, Christiaan and Constantijn, with equipment for making lenses.[123] This gun also was preserved in the repository.[124]

One of the specialist bodies that had been set up on John Beale's initiative in 1664, the 'Mechanical Committee', took a particular interest in artillery, while the patent for various inventions that the society obtained in March 1665 included novel types of guns giving 'speedier and more effectual discharge'.[125] In the same month the King, through William Legge, Lieutenant-General of the Ordnance, granted the society 'a gun' for making experiments. Since this was in the context of the work of a committee for the improvement of artillery, and since Hooke was immediately ordered to draw up a list of experiments to this end, we can be confident that this was a large gun.[126]

There were two series of experiments that would justify including a firearm in the frontispiece. In 1661 Brouncker worked on 'the recoiling of guns', using an adjustable mount or 'engine' to carry a firearm and shooting (in the procedure he found successful) at a thin target close to the muzzle.[127] The experiments took place at Gresham College and then at Whitehall, the latter in the presence of the King. We tend to think of recoil as a motion backwards, but Brouncker was concerned more with lateral displacement and its effect on accurate targeting. The second series of experiments dates from August 1664, when Hooke tried unsuccessfully to measure the velocity of a bullet, at a society meeting, by adapting a pendulum-regulated instrument used for timing falling bodies. The pendulum was to be started by the bullet leaving the mouth of the gun and stopped by the action of a string stretched to the target.[128]

It seems extremely unlikely that either of these sets of experiments used one of the firearms we have identified at the Royal Society. Something as exotic as an airgun or a repeater would surely have been mentioned, and both investigations were more concerned with the actions and properties of firearms in general use. Brouncker reports using more than one gun, and his 'engine' was preserved in the repository with a gun in place, the other two firearms being listed separately.[129] In reporting his experiment, Hooke refers to a 'carabin', or carbine, which would usually indicate a shorter barrel than a common musket.[130]

What can be said, then, of the firearm in the frontispiece? It is certainly not the 'wind-gun', even if airguns might at times have been made to resemble more regular firearms.[131] The physician, collector and experimental philosopher Michael Bernard Valentini of Giessen, who became a Fellow of the Royal Society in 1715, examined the Wilkins gun in the repository and described a firearm with a pump in the butt and a reservoir for the compressed air in the barrel, and he published a diagrammatic sketch incompatible with the gun in the frontispiece.[132] Might it instead be the Kalthoff repeater?

Just as there are no grounds for thinking that a gun recorded at the society was used in any of these trials, we are not compelled to think that the gun in the print either belonged to the society or was used in the documented experiments. It is unique among the instruments in the print in being an example of an item of general commerce. It was not possible simply to purchase an example of Boyle's air-pump or of Hooke's micrometer quadrant in the London instrument trade, although it might have been possible to place a special commission for a bespoke item. Neither, at this date, could you buy a wheel barometer or a sealed thermometer, even though they could perhaps have been had from the society's operator. An obvious absentee from the frontispiece is the compound microscope, perhaps because it was indeed a commercial item, albeit in rather a specialist trade. That the firearm is present is partly because of the experiments at Gresham College (which could, of course, be cited in the case of the microscope), but perhaps also because these experiments were of immediate practical concern to important patrons the society were anxious to cultivate, including Prince Rupert and the King (which could not).

We consulted experts in the history of firearms for help with an identification. One observation in response was familiar: the artist did not always pay attention to including detail that must have been present. The Kalthoff repeater was not favoured and the overall conclusion, communicated by Jonathan Ferguson of the Royal Armouries, following a conversation with his colleague Graeme Rimer, was 'a short carbine or sporting gun with what appears to be a snapping matchlock mechanism'.[133] Graeme Rimer separately proposed the specific suggestion of a matchlock harquebus, 'the shorter and lighter longarm which preceded the arrival of what we would now call the musket'.[134] These were Italian guns imported into England in large numbers from the mid-sixteenth century.[135] What is particularly interesting is that both the general and the specific identifications are consistent with Hooke's reference to a 'carabin', made perhaps because by the 1660s the relative shortness of the gun was a matter of note. While both Brouncker's experiments of 1661 and Hooke's of 1664 helped justify the inclusion of a firearm in the print, once again Hooke may have been responsible for the presence in Gresham College, in the core period we have already identified, of an experimental object that found its way into Sprat's frontispiece.

Puzzles and uncertainties

A few objects have stubbornly resisted identification and, unlike number 15, offer limited clues on which to work. Number 9 may be a simple object in its own right, or may be a pillar stand on a flat, round base, supporting the triangular piece immediately behind it. The latter has the general shape of Wren's self-emptying rain gauge, where the displacement of the centre of gravity as it filled would cause it to tip up at a known capacity and then right

Figure 5.37 Instrument number 9.

itself for refilling.[136] It is understood, however, that this reservoir would have been wider than is shown and certainly the example preserved in the repository, attributed to Wren, was suspended in a frame, rather than supported directly on a pillar.[137]

We have seen that in September 1664 Hooke was ordered to contrive a weather clock 'as well and as cheap as he could', so he might well have begun to work in this area soon afterwards and he was aware of Wren's earlier device.[138] Hooke's account of his own self-emptying rain-gauge, published by William Derham in his *Philosophical Experiments and Observations of the Late Eminent Dr Robert Hooke* (1726), was written much later, in December 1678, when he presents it in the context of his 'Weather-Wiser', or weather-clock.[139] His diagram offers a profile of his 'Triangular Prism' close to the shape in the frontispiece and he describes it as 'poiz'd like a Balance upon a Foot', that is, in the manner as it appears there.[140] If the dot towards the top of the pillar indicates a pivot, it is in the appropriate position on the prism. However, Hooke's account also says that there was a second pivot at 'the like opposite Point in the other Side of the Vessel'; for this to be so, the stand in the print must be a fork mount, which is not impossible, but is not indicated with any clarity.

Immediately below this puzzle, number 10 presents another. A substantial round object (possibly in the shape of a wide-walled cylinder or of a toroid) sits a little precariously at the bottom of an inclined platform, which may be a frame rather than a solid plane. This is set on a low table with turned legs. Is it part of something else, whether present elsewhere in the image or not? Or is it an instrument in its own right? We have considered a number of possibilities and others have been tentatively suggested by obliging colleagues, but we do not have the confidence to advocate any of them here.[141]

Number 14 is perhaps the least communicative object of them all. The triangular shape above represents a wire or cord suspension and its shadow, so we are left simply with a rectangular box shape having an aperture at the bottom, and any hints of further detail are very slight. To judge by the surrounding objects, which is not a reliable strategy, it is perhaps twenty inches high and about nine inches across.

Figure 5.38 Instrument number 10.

Figure 5.39 Instrument number 14.

Here is one suggestion. The society began a discussion of building materials in October 1666, the month following the Great Fire of London. Bricks became the focus of attention and Hooke the node of activity.[142] Towards the end of March 1667 he proposed 'an expeditious way of making bricks' and at subsequent meetings the society urged him to make good on this.[143] On 18 April, 'Mr. Hooke produced his model for brick-making, and promised to produce another at the next meeting.'[144] Whether, after a further reminder, the 'brick-engine' that was 'produced again' on 9 May and tried out was something different is not clear, but it was a real object: the society set about making bricks with clay, 'but that being too stiff, the trial succeeded not'.[145]

Number 14 could be a mould, Hooke's 'model for brick-making'. The aperture at the bottom is about the right size and has the proportions appropriate to a brick. Imagining how that shape of clay was removed from the mould would take speculation too far, but it would have been part of Hooke's 'expeditious way'. If this is right, the mould was probably the last object to be included in the design for the print. From the dating of the drawing of the 60-foot telescope, we know that the plate was not etched before very late in February 1667, and it may not have been completed even as late as the end of April (see chapter 2). What could be a more appropriate addition to the manifesto of a 'Corporation', to use Sprat's terminology, intent on asserting its practical worth in London in 1667?[146]

Conclusion

While not everything has been identified with confidence, we can now characterise the objectives of the print with greater clarity and precision.

We noted at the start of this chapter that the frontispiece does not include rarities, whether natural or artificial; neither does it illustrate more everyday specimens from the natural world. It is not, therefore, a representative selection of the contents of the society's growing repository. We have already seen how in 1666 the society augmented its collection by purchasing the rarities that had been assembled by Robert Hubert, in addition to accepting gifts of birds, snakes, an antelope skin, rocks and minerals in the same year.[147] Wilkins's original gift of 1663 had included an ostrich egg and a coconut.[148] Nothing of the kind is depicted. Perhaps there are too few illustrations of contemporary cabinets to identify any as 'typical' but those we have, such as the famous 1655 print of the cabinet of the Danish physician and collector Ole Worm, are filled with objects to look at.[149] The instruments in the Sprat frontispiece are not for examination and wonder but for action. They are ready for use in practical mathematics and experimental philosophy.

The frontispiece seems to present a cabinet of a different character, a cabinet of experiment rather than curiosity, delivering answers and applications rather than questions and spectacle, a cabinet full of confidence and expectation. This vision would stimulate a great many foundations in the eighteenth century, as the 'cabinet of physics' became the typical social and educational space for experimental philosophy throughout Europe and beyond.[150] It is appropriate that a very early and remarkably precise blueprint for such an institution seems to have struck Sir William Petty at a meeting of the society in December 1673, after witnessing yet another of Hooke's countless experiments, on this occasion on the distance law of magnetic attraction:

> Sir William Petty moved, that the Society would give orders, that there might be a constant apparatus of instruments ready for the making of several kinds of experiments depending on several heads; for instance, for experiments of motion, optical, magnetical, electrical, mercurial, &c. And that such instruments, as had been formerly used by the Society, and were out of order, might be repaired, and all these put together in a room by themselves, to be ready upon occasion for strangers, or for repetition and farther prosecution of the several sorts of experiments.[151]

It is ironic, perhaps, that in spite of this clear-sighted and prescient proposal, and although Hooke's regular demonstrations, in fulfilment of his obligations as Curator of Experiments, were influential for other institutions, the Royal Society never did establish a 'cabinet of physics', as so many other learned corporations would do in the following century. Although the instruments in Sprat's frontispiece were real, their assemblage was not.

Mathematical and, latterly, optical instruments were already an established commercial presence in London, as Sprat himself remarked: 'the ordinary shops of *Mechanicks*, are now as full of *rarities*, as the *Cabinets* of the former *noblest Mathematicians*'.[152] Experimental philosophy was just beginning to appear in these shops and workshops, a feature of the London trade that would become important in the eighteenth century, as makers developed an experimental commerce through instruments, demonstrations, lectures and books.[153] The repository flourished for a time and was accommodated in its own building after the society had moved to Crane Court in 1710.[154] It did contain some instruments, but it is clear from the catalogue published by

Nehemiah Grew in 1681 and the manuscript inventories that survive from the eighteenth century that instruments were a very small minority of the collection.[155] The repository did not become a site for experiment and demonstration. When the German traveller Zacharias Conrad von Uffenbach was seeking such entertainment he found it, not in the repository, which he visited, but a few doors away, in the premises of Hooke's successor as Curator, the instrument maker Francis Hauksbee.[156]

Notes

1 Birch, *Royal Society*, vol. 1, pp. 123, 124, 453, 488, 490, 496, 499, 507, 510; vol. 2, pp. 3, 4. For the background of the tortuous negotiations by the society with Sir John Cutler concerning his endowment of a lectureship for Hooke, see Michael Hunter, 'Science, Technology and Patronage: Robert Hooke and the Cutlerian Lectureship' in Hunter, *Establishing*, ch. 9.

2 Boyle, *Correspondence*, vol. 2, p. 343. On Hooke's Gresham Lectureship see Michael Cooper, 'Hooke's Career', in Jim Bennett, Michael Cooper, Michael Hunter and Lisa Jardine (eds), *London's Leonardo* (Oxford, 2003), ch. 1, esp. pp. 21ff.

3 Birch, *Royal Society*, vol. 2, p. 73.

4 Evelyn, *Diary*, vol. 3, p. 433.

5 Birch, *Royal Society*, vol. 2, p. 96.

6 See M.S. Mahoney, 'Christiaan Huygens: The Measurement of Time and of Longitude at Sea', in H.J.M. Bos, M.J.S. Rudwick, H.A.M. Snelders and R.P.W. Visser (eds), *Studies on Christiaan Huygens* (Lisse, 1980), pp. 234–70, and J.H. Leopold, 'The Longitude Timekeepers of Christiaan Huygens', in W.J.H. Andrewes (ed.), *The Quest for Longitude* (Cambridge, MA, 1996), pp. 102–14.

7 For one such clock, signed by Severyn Oosterwijck, see Keith Piggott's *appendix 5*: http://www.antique-horology.org/piggott/rh/, and Richard Dunn and Rebekah Higgitt, *Ships, Clocks & Stars: The Quest for Longitude* (London, 2014), pp. 57–63, including fig. 18. We are grateful for the very helpful cooperation of John Cheetham, who identified this clock and has generously shared his knowledge with us. We also thank Keith Piggott for his generosity with information and images. Piggott, *appendix 5*, deals also with the unsigned clock, and we are grateful to Richard Dunn for drawing our attention to its presence in the National Maritime Museum, see http://collections.rmg.co.uk/collections/objects/618756.html (accessed 20 December 2015). Note also Lisa Jardine, 'Scientists, Sea Trials and International Espionage: Who Really Invented the Balance Spring Watch', *Antiquarian Horology*, 29 (2006), 663–83.

8 Christiaan Huygens, *Oeuvres complètes* (22 vols., Amsterdam and The Hague, 1888–1950), vol. 4, p. 318. There are rough sketches by Huygens of triangular longitude clocks in vol. 17, p. 165.

9 William Derham (ed.), *Philosophical Experiments and Observations of the Late Eminent Dr. Robert Hooke* (London, 1726), p. 4.

10 Birch, *Royal Society*, vol. 1, pp. 320, 331; vol. 2, pp. 4–5, 23–4, 26; RBO, vol. 2, part i, p. 205.

11 'A Narrative Concerning the Success of Pendulum-Watches at Sea for the Longitudes', *Phil. Trans.*, 1 (1665–6), 13–15.

12 Hunter, *Establishing*, p. 89 (though the item in question is there mistakenly associated only with Hooke, rather than with Bruce and Huygens); see p. 12 in this book.

13 Birch, *Royal Society*, vol. 2, p. 112.

14 Anthony Turner suggests, in a private communication, that there are indeed two hands and that the lower is a minute hand with a feature found on Dutch clocks of the period, namely a 'clover leaf or anchor or laterally spreading ends above the (tiny) pointer', the small scale of the print creating the visible approximation to 'a sort of crescent'. He cites examples in Reinier Plomp, *Spring-Driven Dutch Pendulum Clocks, 1657–1710* (Schiedam, 1979). The slight kink in the depiction would support this suggestion. However the illustrations in Pomp also show that the hour hand was typically more elaborately shaped than the minute, and in the Sprat print the upper end is a simple pointer. Illustrations worth examining for comparison are at pp. 20, 30–1 and esp. 35.

15 Piggott, *appendix 5* (see n. 7).

16 We are very grateful for the help given by Diana Crawforth-Hitchins in appreciating the significance of this instrument.

17 Birch, *Royal Society*, vol. 1, pp. 466–7, 468–9.

18 Ibid., vol. 1, pp. 163–5.

19 Ibid., vol. 1, p. 164.

20 The details that follow are taken from a letter from Hooke to Boyle, 15 September 1664, Boyle, *Correspondence*, vol. 2, pp. 324–5; see also Birch, *Royal Society*, vol. 1, p. 467.

21 Boyle, *Correspondence*, vol. 2, p. 325.

22 Robert Hooke, *Animadversions on the First Part of the Machina Coelestis of the Honourable, Learned, and Deservedly Famous Astronomer Johannes Hevelius* (London, 1674), p. 44.

23 Sprat, *History*, p. 246.

24 Ibid.

25 Charles H. Cotter, *A History of the Navigator's Sextant* (Glasgow, 1983), pp. 104–10; note also Charles H. Cotter, 'The Mariner's Sextant and the Royal Society', *Notes & Records*, 33 (1978), 23–36.

26 Birch, *Royal Society*, vol. 2, p. 111.

27 Ibid., vol. 2, p. 113.

28 Ibid., vol. 2, pp. 113, 114. What Hooke 'presented' (i.e., demonstrated) and is depicted in the frontispiece may have been a wooden model, for listed in the repository in an inventory of c.1730 was, 'a wooden model of Dr. Hook's Reflecting Quadrant', Royal Society Archives, MS 414, fol. 10.

29 Birch, *Royal Society*, vol. 2, p. 115.

30 Robert Hooke, *The Posthumous Works*, ed. Richard Waller (London, 1705), p. 503 and tab. 11, fig. 2. The reader should note that in Fig. 5.7, as also in Figs. 5.9, 5.12, 5.14, 5.29, 5.31 and 5.34, we have extracted individual 'figs.', or occasionally details from them, from composite plates on which a number of such images are juxtaposed.

31 Birch, *Royal Society*, vol. 2, p. 58.

32 Hooke, *Posthumous Works*, pp. 557–8 and tab. 12, fig. 5.

33 Ibid., p. 558.

34 Birch, *Royal Society*, vol. 2, p. 58.

35 Royal Society Archives, MS 414, fol. 8; also MS 417, p. 12 (an inventory of 1765).

36 Birch, *Royal Society*, vol. 2, p. 18.

37 Hooke, *Posthumous Works*, pp. 508–9 and tab. 10, fig. 2.

38 Royal Society Archives, MS 414, fol. 8.

39 Evelyn, *Diary*, vol. 3, p. 285.

40 Birch, *Royal Society*, vol. 2, p. 97.

41 Evelyn, *Diary*, vol. 3, p. 441.

42 Birch, *Royal Society*, vol. 2, p. 105.

43 Ibid., vol. 2, pp. 134, 137, 139, 150–1, 153; Louise Diehl Patterson, 'Pendulums of Wren and Hooke', *Osiris*, 10 (1952), 277–321; Patri J. Pugliese, 'Robert Hooke and the Dynamics of Motion in a Curved Path', in Michael Hunter and Simon Schaffer (eds), *Robert Hooke: New Studies* (Woodbridge, 1989), pp. 181–205.

44 Birch, *Royal Society*, vol. 2, p. 150.

45 Sprat, *History*, p. 247.

46 Hooke, *Animadversions*, pp. 69–70.

47 Huygens, *Oeuvres*, vol. 7, p. 323; *Oldenburg*, vol. 10, p. 67 (we rely on a slightly different understanding of 'et mesmes en fit construire des horologes' from the translation given in *Oldenburg*).

48 RIBA Library, 'Heirloom' copy of Christopher Wren jun. (ed.), *Parentalia, or Memoirs of the Family of the Wrens* (London, 1750), ref. no. VOS/233.

49 Birch, *Royal Society*, vol. 1, p. 341; RBO, vol. 2, part i, pp. 321–2.

50 Birch, *Royal Society*, vol. 1, p. 467; vol. 2, pp. 1, 436; vol. 3, pp. 75, 78 (two reminders), 96–7, 222, 432, 445, 450, 453, 480, 486.

51 Grew, *Musæum*, pp. 357–8; Royal Society Archives, MS 414, fol. 4.

52 Keith Piggott, *A Royal 'Haagseklok': 'Severyn Oosterwijck Haghe Met Privilege'*, appendix 3, http://www.antique-horology.org/piggott/rh/appendix3.pdf (accessed 13 January 2016).

53 William Derham, *The Artificial Clock-Maker. A Treatise of Watch & Clock-work* (London, 1696), pp. 95–6.

54 Birch, *Royal Society*, vol. 1, p. 98 (cf. p. 324).

55 A.D.C. Simpson, 'Richard Reeve – The "English Campani" – and the Origins of the London Telescope-Making Tradition', *Vistas in Astronomy*, 28 (1985), 357–65, on p. 359.

56 See *Oldenburg*, vol. 3, pp. 347–9; Cambridge University Library, MS Add. 9579/13/5, fol. 117.

57 The key letters, with the exception of 'A', have been added to both the text and the sketch. Oldenburg made a fair copy of the section of Hooke's letter describing the mounting, including his own insertions and the sketch with his incorrectly placed key letters, see Cambridge University Library, MS Add. 9579/13/5, fol. 116. This sketch is reproduced relatively accurately in S. J. Rigaud (ed.), *Correspondence of Scientific Men in the Seventeenth Century* (2 vols., Oxford, 1842), vol. 1, plate 2, fig. 2 (facing p. 240; the letter appears in vol. 1, pp. 179–82).

58 See *Oldenburg*, vol. 3, pp. 350–5 and plate 3.

59 Reproduced in A.D.C. Simpson, 'Robert Hooke and Practical Optics', in Hunter and Schaffer (eds), *Robert Hooke: New Studies*, pp. 33–61, plate 5, and in Jim Bennett, 'Hooke's Instruments', in Bennett et al., *London's Leonardo*, pp. 63–104, fig. 31; and discussed in Simpson, pp. 37–9, esp. pp. 38 n. 17 and 39 n. 20, and Bennett, p. 94. The original is at Royal Society Archives, Cl. P. 20, no. 61, fol. 134.

60 Royal Society Archives, Early Letters W.3, no. 2; Cl. P. 20, no. 61. Sprat refers to Wren's tube in 1659: Wren (ed.), *Parentalia, or Memoirs of the Family*, p. 254.

61 Birch, *Royal Society*, vol. 2, p. 98, and esp. Simpson in *Robert Hooke: New Studies*, pp. 39–40, nn. 17, 20.

62 Royal Society Archives, Early Letters W.3, no. 2.

63 Cl. P. 20, no. 61, fols. 129–34.

64 A similar device appears in a portrait of the French chemist, Samuel Cottereaux Duclos (undated, but probably late 1660s), National Library of Medicine, image B04869. These differ from that shown in the frontispiece in that the apparatus shown in the Leclerc and Duclos images is made of copper (hence the

angularity from soldering together mitred copper tubes) rather than glass (which can be easily bent into a curved shape).

65 We wish to thank Lawrence M. Principe for his advice on the chemical equipment. Our account derives from his identifications and commentary.

66 Birch, *Royal Society*, vol. 1, pp. 144, 153, 158–9, 439.

67 *Oldenburg*, vol. 3, p. 45; Boyle, *Correspondence*, vol. 3, p. 79 (our italics).

68 See Boyle, *Works*, vol. 6, esp. pp. 35–8 and plate on p. 26, and vols. 6 and 7, passim. For accounts of the development of the air-pump, see George Wilson, 'On the Early History of the Air-Pump in England', *Edinburgh New Philosophical Journal*, 48 (1849), 330–54, esp. pp. 337–9; Michael Hunter, *Boyle: Between God and Science* (New Haven and London, 2009), pp. 157–9; Steven Shapin and Simon Schaffer, *Leviathan and the Air-Pump: Hobbes, Boyle and the Experimental Life* (2nd edition, Princeton, 2011), pp. 256ff.

69 Shapin and Schaffer, *Leviathan and the Air-Pump*, pp. 256–64.

70 See, for example, Birch, *Royal Society*, vol. 1, p. 287; vol. 2, pp. 20, 22–3, 25, 46.

71 Lorenzo Magalotti, *Saggi di Naturali Esperienze Fatte nell'Accademia del Cimento* (Florence, 1667), pp. 1–11.

72 Eleanor Smith Godfrey, *The Development of English Glassmaking, 1560–1640* (Oxford, 1975), pp. 246–50. We owe this point and the reference to Anthony Turner.

73 John Evelyn, *Elysium Britannicum, or the Royal Gardens*, ed. John E. Ingram (Philadelphia, 2001), p. 251.

74 Gaspar Schott, *Mechanica Hydraulico-Pneumatica* (Frankfurt, 1657), p. 231 and plate XIII, fig. IV.

75 Evelyn, *Elysium Britannicum*, p. 251.

76 British Library Add. MS 78342, fol. 189. The note is jotted in and now partly faded. The printed transcript at Evelyn, *Elysium Britannicum*, p. 251, omits 'way', which is clearly present.

77 Robert Boyle, *New Experiments and Observations Touching Cold* (1665), in *Works*, vol. 4, pp. 246–7.

78 Edmond Halley, 'An Account of Dr Robert Hook's Invention of the Marine Barometer', *Phil. Trans.*, 22 (1700–1), 791–4, on p. 788.

79 Toby Barnard, 'Southwell, Sir Robert (1635–1702)', *Oxford Dictionary of National Biography* (Oxford University Press, 2004), http://ezproxy-prd. bodleian.ox.ac.uk:2167/view/article/26066 (accessed 4 October 2015). For thermometry in the Accademia del Cimento and the early Royal Society, see W. E. Knowles Middleton, *A History of the Thermometer and Its Use in Meteorology* (Baltimore, 1966), pp. 27–39, 41–8; Louise Diehl Patterson, 'The Royal Society's Standard Thermometer, 1663–1709', *Isis*, 44 (1953), 51–64; Louise Diehl Patterson, 'Thermometers of the Early Royal Society', *American Journal of Physics*, 19 (1951), 523–35; Terry Quinn, 'In this Issue', *Notes & Records*, 58 (2004), 245–7.

80 See, for example, http://catalogue.museogalileo.it/object/FiftydegreeThermometer.html (accessed 7 October 2015).

81 Birch, *Royal Society*, vol. 1, p. 156. In October 1662 there was a reference to 'a thermometer of quicksilver' being used to regulate a furnace, p. 11.

82 Ibid., p. 164; note also p. 211.

83 Ibid., p. 300.

84 Ibid., p. 302.

85 Ibid., pp. 322, 345, 380.

86 Ibid., p. 314.

87 Boyle, *Cold*, preface, *Works*, vol. 4, p. 216; Robert Hooke, *Micrographia, or, Some Physiological Descriptions of Minute Bodies* (London, 1665), p. 38.

88 Hooke, *Micrographia*, p. 38.
89 Birch, *Royal Society*, vol. 1, p. 380.
90 Ibid., vol. 2, p. 16.
91 Ibid., vol. 2, p. 206.
92 Ibid., vol. 1, pp. 29–30; Cl. P. 7(1)2.
93 Birch, *Royal Society*, vol. 1, pp. 69, 106, 108, 218; Cl. P. 8(1)5; RBO, vol. 1, p. 29.
94 Birch, *Royal Society*, vol. 1, pp. 68, 78, 86–7, 207, 212, 215, 259, 280, 287, 297, 307; Cl. P. 6/5, Cl. P. 20, no. 8; RBO, vol. 1, pp. 153, 178–9; RBO, vol. 2, part 1, pp. 208–10.
95 Birch, *Royal Society*, vol. 1, p. 307; Cl. P. 20, nos. 23, 35; RBO, vol. 2, part 1, pp. 301–2; RBO, vol. 2, part 2, pp. 202–3.
96 Birch, *Royal Society*, vol. 1, pp. 316, 322, 328, 331, 467; *Oldenburg*, vol. 2, p. 235. There was a sea-sounder in the repository in 1765, Royal Society Archives, MS 417, p. 14. For the king's planned visit see p. 62 in this book.
97 'Directions for Seamen, Bound for Far Voyages', *Phil. Trans.*, 1 (1665–6), 140–3 (no. 8); RBO, vol. 1, pp. 149, 151–2.
98 'An Appendix to the Directions for Seamen, Bound for Far Voyages', *Phil. Trans.*, 1 (1665–6), 147–9 (no. 9).
99 'Directions for Observations and Experiments to be Made by Masters of Ships, Pilots, and Other Fit Persons in Their Sea-Voyages', *Phil. Trans.*, 2 (1666–7), 433–48 (no. 24); Birch, *Royal Society*, vol. 2, p. 163.
100 Birch, *Royal Society*, vol. 1, p. 365.
101 Ibid., vol. 1, pp. 367, 371.
102 Boyle, *Correspondence*, vol. 2, p. 343.
103 Hooke, *Micrographia*, preface and Schem. I, fig. 1.
104 Boyle, *Correspondence*, vol. 3, p. 119.
105 Robert Hooke, 'A New Contrivance of Wheel-Barometer, much more Easy to be Prepared, than that, which is Described in the Micrography', *Phil. Trans.*, 1 (1665–6), 218–19 (no. 13) and fig. 1.
106 Cl. P. 20, no. 32; RBO, vol. 3, p. 2; E. N. da C. Andrade, 'Wilkins Lecture. Robert Hooke', *Proceedings of the Royal Society of London. Series A, Mathematical and Physical Sciences*, 201 (1950), 439–73, on p. 449.
107 Sprat, *History*, p. 173.
108 J. A. Bennett, *The Mathematical Science of Christopher Wren* (Cambridge, 1982), p. 84.
109 Hooke, *Micrographia*, pp. 149–51, and Schem. XV, fig. 4; Sprat, *History*, p. 173.
110 Birch, *Royal Society*, vol. 1, p. 311.
111 Ibid., vol. 1, p. 312. On the hygrometer, see W. E. Knowles Middleton, *Invention of the Meteorological Instruments* (Baltimore, 1969), pp. 90–1. We are grateful to Stuart Talbot for the hygrometer suggestion.
112 In November 1663 John Wilkins gave the society, among other things, scales for weighing 'without any counterpoise', which suggests a spring balance, but it was for weighing gold, which would imply a sensitivity not immediately evident here; Birch, *Royal Society*, vol. 1, p. 324. It survived in the repository at least until c.1730, Royal Society Archives MS 414, fol. 3.
113 The Museo Galileo, Florence, has an odometer attributed to Christoph and Hans Christoph Schissler, late sixteenth century, http://brunelleschi.imss.fi.it/mediciscienze/emed.asp?c=35463 (accessed 2 December 2015). We are grateful to Howard Dawes for the suggestion of an odometer, and to Anthony Turner for that of a pedometer.

114 Norman E. Wright and Jane Insley, 'Odometer', in Robert Bud and Deborah Jane Warner (eds), *Instruments of Science: An Historical Encyclopedia* (New York and London, 1998, pp. 423–4.

115 Bennett, *Mathematical Science of Christopher Wren*, p. 74; Birch, *Royal Society*, vol. 2, pp. 483–4, 492.

116 Birch, *Royal Society*, vol. 1, p. 131; Sprat, *History*, p. 250.

117 Grew, *Musæum*, pp. 360–1.

118 Evelyn, *Diary*, vol. 3, pp. 110, 196.

119 Again a suggestion we owe to Anthony Turner. For the pedometer, see Jane Insley, 'Pedometer', in Bud and Warner, *Encyclopedia*, pp. 440–1.

120 Ibid., pp. 324, 335; Grew, *Musæum*, p. 366; Royal Society Archives, MS 414, fol. 13. It was presumably a different gun that was 'applied to the compressing engine' with some success, at Hooke's suggestion, in 1664, Birch, *Royal Society*, vol. 1, pp. 345, 367, 371.

121 Birch, *Royal Society*, vol. 1, p. 332.

122 Guy M. Wilson, *The Vauxhall Operatory: A Century of Inventions before the Scientific Revolution* (Leeds, c.2009), pp. 8–12.

123 Huib J. Zuidervaart, 'The "Invisible Technician" Made Visible: Telescope Making in the Seventeenth and Early Eighteenth-Century Dutch Republic' in Alison D. Morrison-Low, Sven Dupré, Stephen Johnston and Giorgio Strano (eds), *From Earth-Bound to Satellite: Telescopes, Skills and Networks* (Leiden, 2012), pp. 41–102, on p. 81; Dutch Instrument Makers, online directory, http://www.dwc.knaw.nl/biografie/scientific-instrument-makers/?pagetype=authorDetail&aId=PE00012267 (accessed 23 October 2015).

124 Grew, *Musæum*, p. 366; Royal Society Archives, MS 414, fol. 13.

125 See Sprat, *History*, pp. 233ff.; Hunter, *Establishing*, pp. 88–9, 117–18.

126 See above, p. 96. See also Birch, *Royal Society*, vol. 1, p. 24.

127 Ibid., vol. 1, pp. 8, 12, 16, 20, 33, 74; Sprat, *History*, pp. 233–9; RBO, vol. 1, pp. 143–8.

128 Birch, *Royal Society*, vol. 1, pp. 461, 465, 474; Sprat, *History*, pp. 225, 249. For the instrument for timing falling bodies, see Birch, *Royal Society*, vol. 1, pp. 449, 460, 464, 467.

129 RBO, vol. 1, p. 148 ('Some other Experiments I have also made with another Peese (about the same length, but with a bore neere two tenths of an inch lesse), and order'd in the same manner.'); Grew, *Musæum*, pp. 365–6; Royal Society Archives, MS 414, fol. 13.

130 Sprat, *History*, p. 238; Birch, *Royal Society*, vol. 1, p. 461.

131 For contemporary airguns, see Howard L. Blackmore, *Hunting Weapons* (London, 1971), pp. 315–24. We are indebted to Graeme Rimer for this and other references on the history of firearms.

132 Michael Bernard Valentini, *Museum Museorum* (3 vols., Frankfurt, 1705–14), vol. 3, p. 13 and plate XX, fig. 1. Note also Howard L. Blackmore, *Guns and Rifles of the World* (London, 1965), p. 91; Arne Hoff, *Airguns and Other Pneumatic Arms* (London, 1972), pp. 34–5 and fig. 40. For Valentini, see Arthur MacGregor, *Curiosity and Enlightenment: Collectors and Collections from the Sixteenth to the Nineteenth Century* (New Haven and London, 2007), p. 62.

133 Private communication, Jonathan Ferguson to Michael Hunter, 27 February 2015.

134 Private communication, Graeme Rimer to Jim Bennett, 13 October 2015:

> A main characteristic of harquebuses of this nature is that they have a somewhat unnecessarily refined lock mechanism. In contrast to the simple variety familiar to us from the firearms of the 17th century, where a lever or trigger connected to the serpentine was simply squeezed and this controlled the speed

of the descent of the smouldering matchcord into the pan of the lock, the Brescian harquebuses had a snap-lock, in which the serpentine holding the match was pulled back against a light spring and held in the cocked position by a simple sear. Squeezing a small trigger released the serpentine which was then of course propelled forward by the spring. What I think helps identify the detail in your image is that in snap-matchlocks the serpentine is at the rear of the lockplate and falls forwards, while in most later squeeze-type matchlocks the serpentine is at the forward end of the lockplate and pivots towards the firer.

135 Alexzandra Hildred (ed.), *Weapons of Warre: The Armaments of the Mary Rose* (Portsmouth, 2011), part 3, pp. 540–2, 553.

136 Asit K. Biswas, 'The Automatic Rain-Gauge of Sir Christopher Wren, F.R.S.', *Notes & Records*, 22 (1967), 94–104; Middleton, *Meteorological Instruments*, pp. 155–7. Grew says that Wren's instrument was a 'triangular Tin-Vessel', Grew, *Musæum*, p. 358. See also Sprat, *History*, p. 313. We are grateful to Anna Marie Roos for the suggestion of the rain-gauge.

137 Grew, *Musæum*, p. 358.

138 Birch, *Royal Society*, vol. 3, p. 222. For Wren's earlier work, see vol. 1, p. 74.

139 Hooke, *Philosophical Experiments and Observations*, ed. Derham, pp. 41–7.

140 Ibid., p. 44.

141 One intriguing suggestion has come from Anthony Turner, who mentions (in a private communication) the possibility of an inclined plane clock, whose invention he dates to c.1650. We offer this in a footnote, being unable to find a documented link with the Royal Society for the appropriate date, although Turner cites the slightly later paper, Maurice Wheeler, 'A Letter . . . Concerning a Movement That Measures Time after a Peculiar Manner, with an Account of the Reasons of the Said Motion', *Phil. Trans.*, 14 (1684), 647–65. He points to the suggestion of a chapter ring on the face of the 'wheel' and remarks that Hollar had probably never seen such a clock before. This is echoed by the independent comment of Frances Willmoth, 'Looks to me suspiciously like an object drawn by someone who wasn't at all sure what they were looking at!' (private communication).

142 Birch, *Royal Society*, vol. 2, pp. 117, 118–9.

143 Ibid., vol. 2, pp. 163, 164, 167.

144 Ibid., vol. 2, p. 168.

145 Ibid., vol. 2, pp. 171, 172.

146 Sprat, *History*, p. 52.

147 Hunter, *Establishing*, pp. 126–35; Birch, *Royal Society*, vol. 2, pp. 45, 104, 108, 118, 124, 156.

148 Birch, *Royal Society*, vol. 1, p. 324.

149 Ole Worm, *Museum Wormianum* (Leiden, 1655).

150 Jim Bennett and Sofia Talas (eds), *Cabinets of Experimental Philosophy in Eighteenth-Century Europe* (Leiden, 2013).

151 Birch, *Royal Society*, vol. 3, p. 115.

152 Sprat, *History*, p. 80.

153 James A. Bennett, 'Shopping for Instruments in Paris and London', in P. H. Smith and Paula Findlen (eds), *Merchants and Marvels: Commerce, Science, and Art in Early Modern Europe* (New York and London, 2002), pp. 370–95; John R. Millburn, *Benjamin Martin: Author, Instrument-Maker, and 'Country*

Showman' (Leiden, 1976); John R. Millburn, *Wheelwright of the Heavens: The Life and Work of James Ferguson FRS* (London, 1988); John R. Millburn, *Adams of Fleet Street, Instrument Makers to King George III* (Aldershot, 2000); Larry Stewart, *The Rise of Public Science: Rhetoric, Technology, and Natural Philosophy in Newtonian Britain, 1660–1750* (Cambridge, 1992).

154 J.A. Bennett, 'Wren's Last Building?', *Notes & Records*, 27 (1972), 107–18; Hunter, *Establishing*, ch. 4.

155 Grew, *Musæum*, pp. 357–69; Royal Society Archives, MSS 414, 417.

156 Bennett, 'Shopping', pp. 376–7.

6   The publication and dissemination of the print

Publication history

Let us start by briefly recapitulating the print's likely chronology, certain clues concerning which came to light in the previous chapters. As we saw at the outset, the image was initially conceived in connection with John Beale's plans for 'Lord Bacon's Elogyes &c' in 1665 and it had already reached a fairly finished state – whether as a drawing or as an actual etched plate – by the time Beale learned in April 1667 that Sprat's book was at long last forthcoming, instead suggesting that the print might be transferred to that. The frontispiece is itself dated 1667, and such clues as the exact match between its depiction of a telescope and that sent by Oldenburg to Hevelius in February 1667 strongly suggest that it was actually executed in the early months of that year. It is possibly also significant, as we saw in chapter 4, that Merrett's *Pinax* is included, since he had presented the society with a copy in January 1667 and this is a book that might otherwise not have seemed worth including.[1] In fact, there are no clues that definitively suggest an earlier (or a later) date.

As for when the book by Sprat to which it was prefixed came out, *The History of the Royal Society* was entered in the Stationers' Company Register on 25 July 1667 and copies of it appear to have become available a few weeks later: Samuel Pepys heard of 'several new books coming out', including Sprat's, at the Royal Exchange on 10 August 1667, while we saw in chapter 2 how Beale had received his copy by 11 September at the latest.[2] The ecstatic comments about the book that he sent to Evelyn in his letter of that date were echoed by the Suffolk virtuoso, Nathaniel Fairfax, in a letter to Henry Oldenburg dated 28 September.[3] As for the frontispiece, a *terminus ante quem* for this is provided by a copy of the *History* bound in red morocco now at the Royal Society, the copy of the plate in which is inscribed by Oldenburg: 'Presented to the R. Society from the Author by the hands of Dr John Wilkins', with the date 'Octob. 10. 1667' (Fig. 6.1).[4]

Having said that, we now need to return to the history of the print once it had been produced, and particularly the question of its relationship to Sprat's book as briefly addressed in chapter 1, a little more about which

Figure 6.1 Oldenburg's inscription in the presentation copy of Sprat's *History* given by John Wilkins to the Royal Society on 10 October 1667.

may be said here. First, it is important to stress that many surviving copies of the book lack this etched frontispiece, including a number that appear to be in their original bindings.[5] In fact, as we have already seen, the book has another frontispiece in the form of an engraving of the society's coat of arms as granted in the second charter – which thus makes the point about the institutional status of the society by which such store was set in its early years as effectively as does Hollar's print. This is printed on the verso of the leaf facing the title-page, which bears the signature A1 and on the recto of which is the book's imprimatur (see Fig. 2.1). It is thus integral to the make-up of the book, and, in copies where the etched plate is present, the relationship between the two is often slightly unsatisfactory, with the etched plate frequently facing the engraved coat of arms. This therefore reinforces the point that the book seems originally to have been conceived without the frontispiece, and it is clear that copies were issued without it and that these are perfectly complete nevertheless.[6]

One possibility that needs to be disposed of here is that the absence of the plate from many copies of the book may be simply explained by the fact that the print-run of the plate would have been less than that of the book. It is notorious that type (and illustrations in the form of woodblocks) were more durable than engravings and especially etchings printed from copper plates, the fine lines of which were liable to become worn when much reused. On the other hand, there is clear contemporary evidence that, if well looked after, even etched plates could produce many hundreds of copies. Indeed Stefano Della Bella claimed concerning the frontispiece that he prepared for the 1642 edition of Galileo's *Opere* that 2,500 or 3,000 impressions could be produced before the plate wore down, though in general retouching would be required to achieve this.[7] Hence, although Sprat's *History* is a common book – with a print run of perhaps 1,500 copies – there is no reason to doubt that impressions of the frontispiece could have been provided for all copies if this had been intended.

Turning now to the copies of the book into which the etched plate *was* bound, it seems to have been inserted in more than one way. One variant is provided by choice presentation copies of the book printed on large paper, in which the plate is bound in front of the title-page, facing forwards, its blank verso facing towards the title-page. Such copies include the one at the Royal Society on which the print itself bears the handwritten presentation inscription in the hand of Henry Oldenburg that has already been quoted.[8] It is also thus that the frontispiece appears in not just one but two copies of the book that were presented to King Charles II, which survive in gilt-tooled morocco bindings of the kind executed for the royal library by the bookbinder, Samuel Mearne: the tooling includes a motif of intertwined Cs under crowns on the spine and at the corners of the tooled boards. One of these is now in the British Library, shelfmark 90.d.18 (Fig. 6.2); the other was formerly in the possession of the bibliophile, Howard C. Levis, and it has twice appeared at auction in the United States in recent years.[9] Because all of these are large

Figure 6.2 Tooled morocco binding executed by Samuel Mearne for the copy of Sprat's *History* presented to Charles II for the royal library, now in the British Library.

paper copies of the book, they have a page size which is just large enough to accommodate the Hollar print.[10]

However, even among large-paper copies there is some variety, since one copy now at the Royal Society (in fact, the copy with the largest dimensions known) has the frontispiece facing the title-page, not with its verso towards it.[11] In copies of the book printed on ordinary-sized paper but in contemporary bindings, on the other hand, the plate is dealt with in a number of ways. In some cases it is in the same position as in the large paper copies, sometimes facing towards the title-page, sometimes with its verso towards it. In certain of these copies, the plate has been severely cropped at the top and bottom in order for it just to fit the page height of the book, while the fore-edge of the print has been folded in.[12] In others, the etching has been folded twice, once vertically and once horizontally.[13] In yet other cases, the binder has used his discretion as to where the frontispiece is inserted, so that, instead of appearing between the title-page and signature A1, it appears elsewhere: in one copy it faces page 1 of the book (Fig. 6.3), and in another

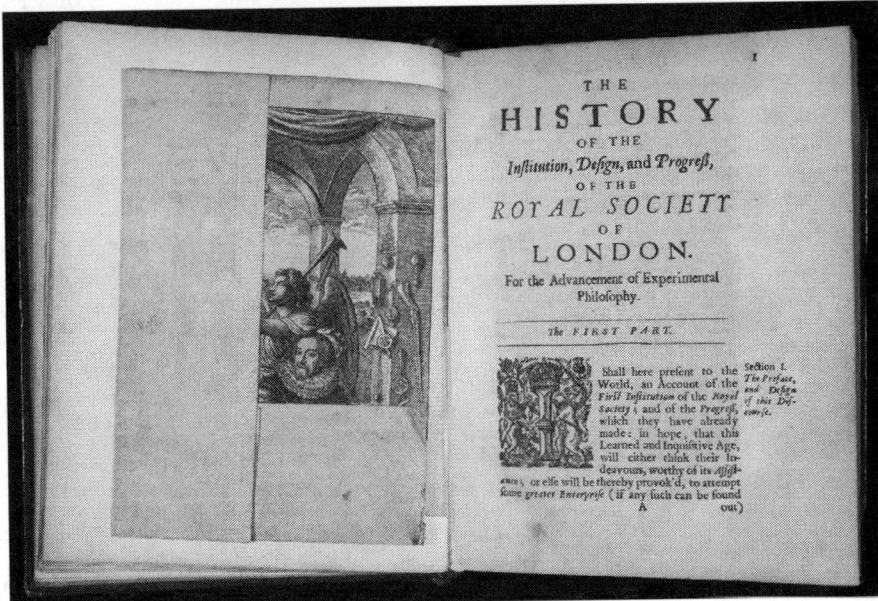

Figure 6.3 An ordinary-sized copy of Sprat's *History* with the print bound in to
face page 1 of the text; in this case, it has been folded twice to fit into
the volume.

it is placed between the Epistle Dedicatory and Abraham Cowley's ode "To
the *Royal Society*".[14]

Beyond this, it is hard to generalise about the relationship between the
book and the print. This is due not least to the fact that the plate has long
been regarded as a necessary part of the volume by collectors and dealers,
meaning that examples of it have often been retrospectively inserted, both
in large paper copies and copies of ordinary size.[15] The earliest instance of a
'made-up' copy of which we are aware is that in the library of Samuel Pepys
at Magdalene College, Cambridge. In this, the copy of the frontispiece is
tipped in before the title-page, but it clearly had a separate history before it
arrived there, as it has a horizontal crease across the centre, having previously been folded in half as a separate print.[16] Pepys's example was followed
by many later collectors, including Thomas Grenville (1755–1846): in his
copy of Sprat's *History*, now in the British Library, the frontispiece is clearly
a retrospective insertion, in this case probably replacing a copy of it that had
been removed.[17]

As the latter instance implies, copies of the print that are today preserved
as loose specimens may well have been removed from copies of Sprat's book,
and at least some of the examples now to be found in the print rooms of

major museums and other collections almost certainly originated thus. For example, three of the five impressions on laid paper in the British Museum are trimmed close to the image and have clearly once been folded for insertion in a book.[18] Other copies survive in 'extra-illustrated' books into which they have been inserted either for the portraits of Brouncker and others that the print includes, or because the print was itself seen as emblematic of the founding of the Royal Society.[19] On the other hand, some of the loose copies that survive have a good-sized margin – including two others at the British Museum – and, though these could have been cut out of large paper copies of the book, there is no reason why impressions of the print should not also have been distributed separately, a common practice at the time.[20] Though it would be gratifying if the print's separate existence could be clinched by finding a copy printed on an original sheet of paper too large ever to have fitted even into a large paper copy of the book, as yet no such copy has been found; in any case, it is not clear what this would prove, since the original copper plate could have been reused to produce such copies at a later date.

The possibility of such reuse is made likelier by the fact that we know that this copper plate survived long after its original deployment in 1667. It seems to have remained in Evelyn's possession after it had been used for the original print run, and it then passed with the rest of his collections to his family. These collections were treasured and even enhanced by Evelyn's Victorian descendant, W. J. Evelyn, and in 1949 they were deposited by J.H.C. Evelyn at Christ Church, Oxford, where they remained for two and a half decades prior to being removed by the Evelyn family: the library, pictures and furniture were then auctioned by Christie's in 1977 and 1978 while the manuscripts were purchased by the British Library in 1995.[21] Among the items that were at Christ Church earlier in the 1970s was a sackcloth bag which bore the following inscription: 'Mr JEvelyns Picture Ingraven by Le Chevalier R Nanteuil At Paris 1650. Also the Frontis-Piece of Dr Sprats Hist. of the R. Society given by Mr Evelyn. This Frontis-piece plate I lately gave to the Society 1733'. The inscription was evidently by Evelyn's grandson, the first baronet.[22] It is perhaps worth noting that the plate of the famous Nanteuil portrait print of Evelyn, which is known to have been in Evelyn's possession, was reprinted in 1818 to illustrate William Bray's first publication of Evelyn's *Diary* in his *Memoirs Illustrative of the Life and Writings of John Evelyn*, though it has since been lost.[23]

In any case, neither the Nanteuil plate nor the Hollar one was in the cloth bag when this was at Christ Church (though some other plates were, including that of one of the landscape etchings by Evelyn referred to in chapter 3). What matters here is the evidence that this provides concerning the fate of the Hollar plate; there is no reason to doubt that Sir John Evelyn, who had himself been elected a Fellow in January 1723, gave this to the Royal Society in 1733, as he states in his inscription on the bag, though there is no record of this in the society's minutes.[24] Its subsequent fate is unclear: ultimately, it is almost certain that, as with other old plates in the society's possession, including those used in early issues of *Philosophical Transactions*, it was sacrificed to the war effort during the First World War.[25] However, prior to that, either when in the possession of the Evelyn family or of the society, it could easily have been used to run off extra impressions of the frontispiece, and any on larger sheets of paper could as easily reflect such retrospective connoisseurship as tell us anything about how the print was originally issued.

Special copies of the print

However, there is clear evidence that at least some choice copies of the print *were* produced at the time when it was etched and printed, and this is as follows. Among the copies of it which survive in the Department of Prints and Drawings at the British Museum are three that Simon Turner, in his definitive catalogue of Hollar's oeuvre for the *New Hollstein* series, singles out as being printed on special paper, two on 'tissue-thin' paper and one on 'golden paper'; a further copy on tissue-thin paper survives at Prague.[26] When examined, the examples of the print among those in the British Museum collection that Turner singled out as being on tissue-thin paper do indeed prove to be impressions on paper that appears to be oriental, more particularly Japanese. These have been examined on a light box and under magnification, and they show all the characteristics of paper produced from so-called *gampi* fibres, possibly mixed with *kozo* fibres, another variety of Japanese paper. This is not only extraordinarily thin and delicate but has a shiny surface and a kind of translucent texture, and impressions made on such oriental paper must almost certainly have been produced as dedication copies or separate collector's items; it should be added that it would have been very difficult to bind them into a copy of the book without mounting them on a further sheet of paper. However, due to its slightly glowing hue and its finely textured and shiny surface, the use of such paper results in superb impressions of the print. For comparison, Hollar's contemporary, Rembrandt, frequently used a thick, lustrous *gampi* in order to create superior impressions of early states of most of the etchings that he produced between c.1647 and 1665.[27] The third relevant example of the print in the Department of Prints and Drawings (the one described by Simon Turner as being on 'golden paper') is on a different stock of paper that is slightly thicker in texture, but this, too, has a distinctive, smooth feel to it, producing a very good quality impression; this also seems to be on *gampi* paper.[28]

It is perhaps worth noting that one of the specimens on tissue-thin paper, P,6.121, has a crease which is very noticeable when the print is illuminated from behind (see Fig. 6.4). It appears that this creasing took place during printing, probably due to the unfamiliarity of the printer with the material that he was using. In other words, it seems likely that, in order to print such copies on the standard rolling press used for intaglio printing, a backing sheet of thicker European paper would have been placed over the Japanese paper, which had the unintended effect of causing it to buckle, meaning

Figure 6.4 One of the British Museum copies of the print on *gampi* paper imported from Japan, photographed over a light box and thus showing the severe crease evidently caused during printing.

that, when it was flattened while passing through the rolling press, a crease was produced from the top to the bottom of the image. In this instance, the crease is so severe that it is perhaps surprising that the impression was not discarded. On the other hand, the quality of other parts of the image is very high due to the superior inking which the *gampi* paper made possible, and this perhaps compensated for the flaw, on balance justifying its retention despite this obvious defect. Nevertheless, the implication is that the paper was too precious to discard even after a mishap like this.[29]

Now it is just conceivable that these copies on special paper are examples of impressions of the print made from the plate at a later date. But it is much more likely that they were made in the 1660s, since Japanese paper is known to have been available in Europe at this time, although not very commonly:

this is shown by Rembrandt's use of it, as already noted, though otherwise our knowledge of such imports is disappointingly sketchy.[30] Here, what is fascinating is that there is evidence that Evelyn knew about such paper and attached a high value to it. He had in fact commented on some paper made from plant fibres that he saw in the collection of Ferrante Imperato at Naples during his Italian travels in 1645, but more directly relevant is a passage in his *Diary* for 22 June 1664 in which he noted how a Jesuit named Tomson showed him 'a Collection of rarities, sent from the *Jesuites* of *Japan* & *China*'. These were destined as a present for the order in Paris, 'to be reserved in their *Chimelium*' (by which Evelyn meant their treasury), but they were brought to London because they came in East India Company ships. Evelyn expostulated how they were such rarities 'as in my life I had not seene', and among them was

> A sort of paper very broad thin, & fine like abortive parchment, & exquisitely polished, of an amber yellow, exceeding glorious & pretty to looke on, & seeming to be like that which my L: *Verulame* describes in his *Nova Atlantis*; with severall other sorts of papers, some written, others Printed.[31]

This extraordinary passage deserves close attention in helping to understand the special examples of the Sprat frontispiece. 'Abortive parchment' is of course parchment made from the skin of stillborn calves and therefore of more than usual subtlety and translucence. It sounds as if Evelyn was shown paper similar to that used for the prints on *gampi* or *kozo* paper, and he was clearly deeply impressed by it. What is equally interesting is the fact that this immediately made him think of Lord Bacon's posthumously published utopian sketch, *New Atlantis* (1627), where Bacon does indeed refer to material 'somewhat yellower then our Parchment, and shining like the Leaves of Writing Tables, but otherwise soft and flexible': this is his description of the parchment scroll that the travellers were given on their arrival at New Atlantis.[32] This therefore suggests, first, that Evelyn became acquainted with oriental paper just at the time when the frontispiece was being conceived, and, second, that he saw this as linked to the technological aspirations expressed by Bacon to which the Royal Society was the heir.

The significance of this will become apparent by recapitulating the background. As is well-known, Bacon saw it as axiomatic that improved understanding of nature should lead to the amelioration of human life: this ambition is found throughout his writings but it perhaps reached its climax in his *New Atlantis*, which provides a breathtaking description of the technological marvels associated with Solomon's House, the publicly funded research institution that he visualised. To this end, Bacon advocated that the study of technology should be central to intellectual enterprise, and it was under his direct influence that the 'history of trades', devoted to the improvement of technical practices, became central to the preoccupations of savants like Evelyn in the 1650s and to the Royal Society after its foundation in 1660, as we saw in chapter 3. In the 1650s Evelyn had produced a volume devoted to '*Trades*. Seacrets & Receipts Mechanical', and both before and after 1660 he seems to have seen it as appropriate to devote his own efforts in this enterprise predominantly to 'Polite' and 'Exotick' arts and trades of the kind that interested virtuosi like himself – mixing colours for painting, for instance, or preparing a superior gilt varnish for picture frames, or making casts of statues.[33] Evelyn's contributions to the Royal Society's history of trades programme in its early years were largely concerned with similar topics, including his account of engraving, *Sculptura*, as noted in chapter 3. Indeed, it clearly seemed only natural to Evelyn that such concerns on his part should be an integral part of the Royal Society's enterprise, thus illustrating that unity of science and art that we explored when examining the link between the design of the frontispiece and the legacy of Raphael.

In fact, as Linda Levy Peck has noted, luxury trades such as those that Evelyn studied were highly valued at the time and were of key economic significance; it is therefore not surprising that they were as prominent in the Royal Society's history of trades programme as they were.[34] Moreover, though the main thrust of the programme was to collect and disseminate information about craft practices to be found in Britain, and to explore how these might be improved by making applied use of the scientific findings of the society's Fellows, both Bacon and his successors were also curious about superior technological practices to be found in other cultures overseas, hoping that these might be emulated at home.[35] It is thus revealing that Evelyn was curious about objects from the Far East not least because they seemed to show superior technical skills unparalleled at home: this is illustrated by another of his comments on the rarities that he was shown by the Jesuit Tomson in June 1664, which included 'Glorious Vests, wrought & embrodered on cloth of Gold, but with such lively colours, as for splendor & vividnesse we have nothing in Europe approches'.[36]

It is into this context that Evelyn's interest in Japanese paper fits. It would surely have seemed appropriate to him to have choice copies of the print that he had prepared to celebrate Bacon and the Royal Society printed on exquisite paper of this kind, no doubt for the benefit of other virtuosi like himself. Moreover, one should not underestimate the degree of connoisseurship involved, since it seems likely that such paper was extremely scarce: only a few Dutch print-makers, mainly apprentices and friends of Rembrandt, occasionally used oriental paper for their prints and drawings.[37] It therefore seems likely that, if it was indeed Evelyn who used *gampi* or *gampi/kozo* paper in this way, he was taking a truly unusual step.

Here, however, a revealing analogy is to be found in Evelyn's *Sculptura*. As we saw in chapter 3, in addition to giving much information about the history of prints for the benefit of collectors, this book also included the earliest published account of the novel art of mezzotint which Evelyn had received from its English pioneer, Prince Rupert.[38] Moreover, Evelyn went so

Figure 6.5 Prince Rupert's mezzotint of *The Little Executioner* as bound into copies of Evelyn's *Sculptura* (1662): this copy, now in the British Library, is
Evelyn's own.

far as to illustrate this new rarity by including a special print in every copy
of the book, which is twice the page size of the book and which is therefore
often bound as a fold-out plate (Fig. 6.5). The illustration in question was
Prince Rupert's mezzotint of the so-called *The Little Executioner* (derived
from a painting then thought to be by the Spanish artist, Jusepe de Ribera),
'which he was pleas'd to honour this *Work* withall, not as a Venal addition
to the price of the Book (though for which alone it is most valuable) but
a particular grace'.[39] In other words, it was included for the benefit of
connoisseurs, and it is symptomatic that, though provided 'as a *Specimen* of
what we have alledged', Evelyn protected the secret of how such prints were
made by failing to divulge key pieces of information. The special impressions
of the Sprat frontispiece on oriental paper in the British Museum and at
Prague are almost certainly further examples of such connoisseurial activity
on Evelyn's part; yet, as his allusion to Bacon in the *Diary* in connection
with paper of the kind used in them shows, it can also be construed as
integral to his commitment to the enterprise of the Royal Society. Such items
simultaneously functioned as scientific objects and as works of art.[40]

It is even conceivable that the use of virtually transparent paper for certain
copies of the print takes us back once again to Raphael. In his *Lives of the
Most Excellent Painters, Sculptors and Architects*, Giorgio Vasari tells how

in 1515 Albrecht Dürer sent to Raphael as a mark of his esteem for him a self-portrait executed on transparent cambric, something which Raphael considered 'a wonderful work'.[41] Could Evelyn have known this story, and was his use of translucent paper for choice copies of the print that he had designed intended to allude to it? Sadly, we will never know, but once again this reminds us of the broader Renaissance context to which this study has shown that the Sprat frontispiece belongs.

Notes

1 See pp. 71, 97–8 in this book.

2 G.E.B. Eyre and C.R. Rivington (eds), *A Transcript of the Registers of the Worshipful Company of Stationers: From 1640 to 1708* (3 vols., London, 1913–14), vol. 2, p. 379; Samuel Pepys, *Diary*, ed. Robert Latham and William Matthews (11 vols., London, 1970–83), vol. 8, p. 380 (though he put the author's name down wrongly as 'Pratt'); he bespoke his copy on 16 August (ibid., vol. 8, p. 387).

3 See p. 19 in this book; *Oldenburg*, vol. 3, pp. 491–3.

4 Royal Society RCN 7384.

5 Examples (limited to copies in contemporary bindings that have not been rebacked or otherwise tampered with) are the copy in the Edward Worth Library, Dublin, illustrated in Fig. 2.1, acquired by John Worth in 1699; or those at Christ Church, Oxford; Corpus Christi College, Oxford (LC 3.b.1); the Countway Library, Harvard; or the Lilly Library, Bloomington, Indiana. (Throughout such listings classmarks will only be given where there is more than one copy of the work in the same library.) Indeed, Allan Chapman in the caption to plate 1 in his *England's Leonardo: Robert Hooke and the Seventeenth-Century Scientific Revolution* (Bristol, 2005) (needless to say, a reproduction of the frontispiece) goes so far as to comment on it: 'while it was popularly said to be associated with Thomas Sprat's *History of the Royal Society* (1667), I have still to find a copy of that work with the engraving bound in'. However, see later in this chapter for copies to which it is undoubtedly integral.

6 Since the book has one sheet in two states (the word 'of' was erroneously repeated in ll. 6–7 on p. 85, and this is corrected in many copies), it could be concluded that the plate might only have been included in copies with the second state of the sheet. In fact, however, it seems likely that this was a stop-press correction, and the incidence of the plate in relation to it seems random: for instance, of the ordinary paper copies of the book in contemporary bindings that lack the frontispiece listed in n. 5 most have the first state, but that in the Countway Library has the second. On the other hand, all large paper copies that we have seen *do* have the corrected version of this page – as might be expected as it was normal practice at the time to print large-paper copies last.

It is perhaps worth noting here that no subsequent edition of Sprat's *History* included the Evelyn/Hollar plate, in other words the French translation (Geneva, 1669; reprinted Paris, 1670), the second edition of 1702, the third of 1722 or the fourth of 1734. However, the retrospective presumption that the book *should* have the plate has affected certain copies of these, too: a copy of the second edition at the National Maritime Museum with a copy of the etching inserted is illustrated in Margarette Lincoln (ed.), *Samuel Pepys: Plague, Fire, Revolution* (London, 2015), p. 204.

7 See Jaco Rutgers, 'A Frontispiece for Galileo's *Opere*: Pietro Anichini and Stefano della Bella', *Print Quarterly*, 29 (2012), 3–12, on p. 9. See also Ad Stijnman, *Engraving and Etching 1400–2000: A History of the Development of Manual*

Intaglio Printmaking Processes (London, 2012), p. 333. It is perhaps worth noting that we have been on the lookout for copies of the print that are noticeably worn, without success; we have also failed to find any evidence of retouching. For the estimate of the likely print run we are indebted to Giles Mandelbrote.

8 Royal Society RCN 7384, bound in tooled red morocco; page size: 238 × 175 mm. Like other large paper copies, this has the second state of p. 85. It is possibly significant that in this copy sig. A1 is missing apart from its stub.

9 It is described in H. C. Levis, *Extracts from the Diaries and Correspondence of John Evelyn and Samuel Pepys Relating to Engraving* (London, 1915), p. 141, and illustrated as the frontispiece and plate 32 to the book. This copy was sold in the Hartz sale at Sotheby's, New York, on 12 December 1991, lot 50, and again in the Pirie sale on 2–4 December 2015, lot 83 (see http://www.sothebys.com/en/auctions/ecatalogue/2015/property-collection-robert-s-pirie-books-manuscripts-n09391/lot.83.html). Before Levis, the copy in question belonged c. 1700 to Richard Graham (fl. 1680–1720), author of the 'short account of the most eminent painters, both ancient and modern' attached to Dryden's translation of Du Fresnoy's *De Arte Graphica* (1695). (We are indebted to Robert Harding for this information.) Its dimensions are 235 × 178 mm; the plate faces forwards.

10 The dimensions of copies of the book are as follows, though obviously allowance has to be made for the activity of binders: ordinary paper, 200 × 155 mm; large paper, ranging from 232 × 172 mm (BL 90.d.18) to 242 × 182 mm (Royal Society COOO 7769: this is due to the fact that the book has not been cropped and retains its deckle edges). The size of the plate mark of the print is 213 × 167 mm.

For Evelyn's own copy of Sprat, which was sadly stolen when on show at Christie's in 1978, see Christie's catalogue, *The Evelyn Library, Part III: M–S*, 15–16 March 1978, lot 1405, with the frontispiece illustrated as plate 24. Its dimensions were 236 × 178 mm, and it was in a typical 'Evelyn' binding; the plate faced forward. For Evelyn's annotations to the book, see the transcript in Hunter, *Establishing*, pp. 69–71.

11 COOO 7769. On the title-page is the signature of Edward Green. The binding of this copy has the arms of Jeanne-Antoinette Poisson, Marquise de Pompadour (1722–64), according to a note by Zaehnsdorf dated 27 October 1937 (presumably when the volume was rebacked).The plate in this copy is not as well inked as normal. For large paper copies which lack the plate, see n. 17 in this chapter. However, in each case it seems likely that the plate was originally present, and therefore that all large-paper copies did include it.

12 For examples with the plate facing the title-page, see e.g., University of London Bromhead Library; Houghton Library, Harvard, *EC65 Sp763 667h. For examples with the plate facing forward, see e.g., Bodleian Library, Oxford, Ashmole 1649; British Library 740.c.17; Cambridge University Library Pet. G. 20 (in this copy, sig. A1 is missing). However, there is reason to believe that the orientation of the print may sometimes have been altered in this and modern times, as is also the case with the pattern of folds in the copies listed in this and the next note.

13 E.g., Cambridge University Library Bassingbourn 321; Hunter c. 66. 6 (facing forwards; the altered pattern of folds is particularly evident in this copy); Yale Medical Historical Library (this initially had a third fold at the bottom which is no longer used); Herzog August Library, Wolfenbüttel; Baillieu Library, University of Melbourne (this measures 223 × 168 mm and is therefore almost a large-paper copy; the plate faces forward).

14 E.g., Cambridge University Library Keynes R.4.31, between Ep. Ded and Cowley's ode; H. B. Wheatley copy (see in this book, Fig. 6.3 and ch. 1, n. 9), facing p. 1.

15 For an example where the plate has been inserted in a large paper copy see n. 17 in this chapter (though matters are complicated by the fact that this seems to replace a copy that had been removed). For another copy of the book in which the plate is smaller than the size of the pages (which is 216 × 165 mm) and is obviously a later insertion, still bearing what are evidently the markings of the eighteenth-century dealer from whom it was bought, see Royal Society RCN 7770. The noticeable disparity of size in this instance obviously does not arise in the case of regular-sized copies of the book.

16 Pepys Library 1529. Pepys also owned a loose copy of the print which is pasted into his album of portraits of 'Gentlemen &c', 2729/124a.

17 See British Library G19477: this is a large paper copy, but the copy of the frontispiece is trimmed to the edge of the design and must be a later insertion. On the other hand, between sigs. A3 and A4 is a full-length stub, presumably of a copy of the print that has since been removed. Another large paper copy from which the plate has clearly been removed is Cambridge University Library Rel.c.66.8 (where a stub remains adjacent to the title-page): see Keynes, *John Evelyn*, p. 284.

18 These are P,5.117, 1859,1008.7 and G,12.168.

19 For details of many copies in print collections see *New Hollstein*, no. 1966. One further copy at the British Museum which is not noted there, G,12.168, is bound into J.C. Crowle's copy of Thomas Pennant's *Some Account of London* (1793), and thus exemplifies the use of the print for extra-illustration. For a further example of this see the copy in the Bull Granger at the Huntington Library, San Marino, vol. 16, no. 22. Note also the copy in Pepys's collection of portrait prints: see n. 16 in this chapter.

20 The copies in the question at the British Museum are P,5.28 (reproduced as the frontispiece to this book) and Q,6.541. For gift-giving of prints at a slightly earlier period, see Michael Bury, *The Print in Italy 1550–1620* (London, 2001), p. 78, and Robert Harding, 'Wenceslaus Hollar and the Earl of Arundel's "Design to Make a Large Volumn of Prints of All His Pictures Drawings & Other Rarities"', in A. Bubenik and A. Thackray (eds), *Perspectives on the Art of Wenceslaus Hollar (1607–1677)* (Turnhout, forthcoming).

21 For information see the essays in *John Evelyn in the British Library* (London, 1995), originally published in *The Book Collector*, 44 (1995).

22 See Edward Gregg, 'Sir John Evelyn: His Grandfather's Heir', in Frances Harris and Michael Hunter (eds), *John Evelyn and His Milieu* (London, 2003), pp. 267–79.

23 For this information see Antony Griffiths, *The Print in Stuart Britain, 1603–89* (London, 1998), p. 132 (catalogue entry 81). The content of the bag when Michael Hunter saw it in the early 1970s was as follows:

> The original plate for JE's print of 'Putney ad Ripam Tamesis'.
> An unfinished plate, with a design of putti based on the pen drawing on the cover sheet of George Evelyn's letter to JE, 4 October 1653 (Add. MS 78303, fol. 70v).
> A plate with a sketch at one corner of a man, a horse and a ruin and shrubbery, inscribed: 'Richardus' and a few other scribbles.
> Two almost wholly blank plates.

These items were unknown to Antony Griffiths when he wrote 'The Etchings of John Evelyn', in David Howarth (ed.), *Art and Patronage in the Stuart Courts: Essays in Honour of Sir Oliver Millar* (Cambridge, 1993), pp. 51–67 (see ch. 3, n. 22 in this book), though on p. 65 (catalogue no. 13), he refers to the 'View of Putney' and to a footnote reference in vol. 1 of Bray's 1818 edition of Evelyn's *Memoirs* which states: 'The plate is now at Wotton'.

24 In view of Sir John's reference to 'lately' in relation to the date 1733, we have searched the society's minutes and the Council minutes from 1731 to 1734, but without success. It is perhaps of interest that, under the date 27 June 1732, the latter refers to the possibility that the society might purchase the plates to Grew's *Anatomy of Plants* and *Musæum Regalis Societatis*: whether anything came of this is unclear, but, even if it did, these too would have been disposed of with the *Philosophical Transactions* plates (see next note).

25 See Sachiko Kusukawa, 'Picturing Knowledge in the Early Royal Society: The Examples of Richard Waller and Henry Hunt', *Notes & Records*, 65 (2011), 273–94, n. 90.

26 *New Hollstein*, no. 1966. The examples in question are 1853,0112.2058 and P,6.121 (tissue-thin paper); and Q,5.640 (golden paper). The example on tissue-thin paper at Prague is R50504. We have not seen this, but Dr Alena Volrábová kindly confirms that the texture of the paper is similar to that of the British Museum specimens: 'It is very thin, but relatively solid, with regular translucent thin stripes. The surface is "silky" – not really shiny, but very fine and smooth' (private communication to Michael Hunter, 20 November 2015).

27 For helpful accounts see George Biörklund and Osbert H. Barnard, *Rembrandt's Etchings: True and False* (revised edition, Stockholm, 1968), pp. 165–79; Jacobus van Breda, 'Rembrandt Etchings on Oriental Papers: Papers in the Collection of the National Gallery of Victoria', *Art Bulletin of Victoria*, 38 (1997), 25–38; Stijnman, *Engraving and Etching*, pp. 262–3, 376–7; and the exhibition catalogues, *Rembrandt: Experimental Etcher* (Greenwich, CT, 1969), esp. pp. 178–80, and *Rembrandt & Saskia: Prints by Rembrandt van Rijn (1606–69)* (Fitzwilliam Museum, Cambridge, 2006), esp. p. 3 (available at http://www.fitzmuseum.cam. ac.uk/dept/pdp/onlinepublications/rembrandt/RembrandtSaskiaHandlist.pdf).

28 This is Q,5.640. Our thanks to Sheila O'Connell for her helpful comment on the 'odd texture' of the paper on which it is printed, as also to Megumi Mizumura for introducing Michael Hunter to the mysteries of *gampi* and *kozo* paper.

29 It is, of course, known that plate printers sometimes needed to retain poor impressions in order to complete an edition: for an example, see Stijnman, *Engraving and Etching*, p. 4, fig. 2. In this instance, it seems likely that a finite quantity of the paper had been obtained and none could be wasted. The superior inking is particularly well illustrated by the figure of Bacon, whose hair on this impression lacks a bald patch that appears on virtually all other copies: we have therefore reproduced this detail in Fig. 4.3a.

30 See n. 27 in this chapter. It is perhaps worth noting that a book printed on *gampi* paper by the Jesuits in Japan in 1590 formed part of the bequest of John Selden to the Bodleian Library, Oxford, and is still there today, Arch. B f.69. In addition, the Bodleian had various Chinese books for which similar paper must also have been used: for a listing, see http://www.bodley.ox.ac.uk/ users/djh/17thcent/17theu.htm, while on Selden's library and its fate see G. J. Toomer, *John Selden: A Life in Scholarship* (2 vols., Oxford, 2009), vol. 2, pp. 793–9. Evelyn was acquainted with John Vaughan, one of Selden's executors, and he visited the Bodleian in 1654 and in 1665 (see Evelyn, *Diary*, vol. 3, pp. 106–8, 385, 492–3), so it is conceivable that he could have learned of these items. For other prints by Hollar printed on tissue-thin paper see *New Hollstein*, nos. 1070, 1071, 1073 and 1076 (all dating from 1649–50): however an examination of the example of no. 1070 in the British Museum, E, 2.85, suggests that it is not on as fine and flexible a paper as the examples of the Sprat frontispiece.

31 Evelyn, *Diary*, vol. 2, p. 331, vol. 3, pp. 373–4. We are grateful to Ad Stijnman for initially drawing the latter reference to our attention (see also Stijnman,

Engraving and Etching, p. 377 n. 57, though he unfortunately failed to identify the Bacon reference cited in the next note); the passage in question is also quoted in *Rembrandt & Saskia*, p. 3

32 Bacon, *New Atlantis* (London, 1627), p. 2. For 'writing tables' comprising reusable leaves of asses' skin, waxed paper or vellum, see Peter Stallybrass, Roger Chartier, J. Franklin Mowery and Heather Wolfe, 'Hamlet's Tables and the Technologies of Writing in Renaissance England', *Shakespeare Quarterly*, 55 (2004), 379–419 (where this passage from Bacon is quoted on p. 390), and H. R. Woudhuysen, 'Writing-Tables and Table-Books', *eBLJ*, 2004, article 3. For Beale's and the Royal Society's interest in parchment-making, including 'a kind of parchment as transparent as glass', see Birch, *Royal Society*, vol. 1, pp. 426, 428; see also John Beale, 'The Art of making Parchment, Vellum, Glue &c. Dec. 14. 1663', Cl. P. 3(i), no. 18.

33 See Michael Hunter, 'John Evelyn in the 1650s: A Virtuoso in Quest of a Role', in Hunter, *Science and the Shape of Orthodoxy: Intellectual Change in Late Seventeenth-Century Britain* (Woodbridge, 1995), pp. 67–98, on pp. 75ff. See also p. 30 in this book.

34 See Linda Levy Peck, *Consuming Splendor: Society and Culture in Seventeenth-Century England* (Cambridge, 2005), ch. 8.

35 An analogy is to be found in the case of Boyle: see Michael Hunter, *Boyle Studies: Aspects of the Life and Thought of Robert Boyle (1627–91)* (Farnham, 2015), esp. pp. 74, 191–2.

36 Evelyn, *Diary*, vol. 3, p. 373.

37 See Biörklund and Barnard, *Rembrandt's Etchings*, p. 173; Stijnman, *Engraving and Etching*, p. 262.

38 See p. 31 in this book.

39 See C. F. Bell (ed.), *Evelyn's Sculptura: With the Unpublished Second Part* (Oxford, 1906), p. 147 and pp. 145ff., passim. A further example of such practice on Evelyn's part is provided by the copies before lettering of the Nanteuil portrait print of him which were evidently intended for presentation on his part: see Griffiths, *Print in Stuart Britain*, p. 132 (catalogue entry 81).

40 For a comparable enterprise in which scientific aspiration and artistic skills were combined, see the study of the making of Martin Lister's *Historiae Conchiliorum* made possible by the discovery at Oxford both of the original plates for the work and of many ancillary proofs and drawings: Anna Marie Roos, 'The Art of Science: A "Rediscovery" of the Lister Copperplates', *Notes & Records*, 66 (2012), 19–40, and 'A Discovery of Martin Lister Ephemera: The Construction of Early Modern Science Texts', *Bodleian Library Record*, 26 (2013), 123–35.

41 Joseph L. Koerner, *The Moment of Self-Portraiture in German Renaissance Art* (Chicago, 1993), pp. 95–6. We are indebted to Matthew Hunter for this reference.

7　Conclusion

The analysis of the frontispiece to Sprat's *History* has thus led us on an extraordinary voyage of discovery, throwing light on many aspects of the early Royal Society and its milieu and not least on the interests and aspirations of John Evelyn. In the course of it, we have been led down avenues of investigation in which we would never have expected to find ourselves, and have benefited from the advice of an extraordinary range of experts in a variety of fields, whose names are recorded in our notes and acknowledgements and to all of whom we are immensely grateful. The overall outcome of the enterprise recorded in this book may be summarised as follows. Evelyn rose to the challenge of John Beale's request for a plate to illustrate his celebration of Lord Bacon and his legacy, producing a powerful composition which, when rerouted to Sprat's *History*, has been influential ever since. An exploration of his sources has revealed how directly Evelyn's overall design followed an exemplar which linked the Royal Society to the legacy of Raphael and of Renaissance Italy. Yet Evelyn adapted this in ways which illustrate much about what he and his Royal Society colleagues considered important about the enterprise in which they were involved, including the institutional status of the new society and the profuse paraphernalia of instruments by which its leading figures set such store. Indeed, the systematic account that has been given of these instruments reveals how closely they reflect the inventions and experiments associated with the society during the years when the frontispiece was being conceived and executed, and particularly the role in devising them of the society's Curator of Experiments, Robert Hooke. Lastly, the way in which copies of the print were evidently printed on special paper itself throws light on the connoisseurial aspirations which formed so important a part of Evelyn's cultural role, yet which were integral to his commitment to the Royal Society's enterprise.

Index